EAI/Springer Innovations in Communication and Computing

Series editor
Imrich Chlamtac, European Alliance for Innovation, Ghent, Belgium

Editor's Note

The impact of information technologies is creating a new world yet not fully understood. The extent and speed of economic, life style and social changes already perceived in everyday life is hard to estimate without understanding the technological driving forces behind it. This series presents contributed volumes featuring the latest research and development in the various information engineering technologies that play a key role in this process.

The range of topics, focusing primarily on communications and computing engineering include, but are not limited to, wireless networks; mobile communication; design and learning; gaming; interaction; e-health and pervasive healthcare; energy management; smart grids; internet of things; cognitive radio networks; computation; cloud computing; ubiquitous connectivity, and in mode general smart living, smart cities, Internet of Things and more. The series publishes a combination of expanded papers selected from hosted and sponsored European Alliance for Innovation (EAI) conferences that present cutting edge, global research as well as provide new perspectives on traditional related engineering fields. This content, complemented with open calls for contribution of book titles and individual chapters, together maintain Springer's and EAI's high standards of academic excellence. The audience for the books consists of researchers, industry professionals, advanced level students as well as practitioners in related fields of activity include information and communication specialists, security experts, economists, urban planners, doctors, and in general representatives in all those walks of life affected ad contributing to the information revolution.

The series has no official board, but EAI has a board it can rely on for feedback https://eai.eu/#!/about/leadership

This series is indexed in Scopus.

About EAI

EAI is a grassroots member organization initiated through cooperation between businesses, public, private and government organizations to address the global challenges of Europe's future competitiveness and link the European Research community with its counterparts around the globe. EAI reaches out to hundreds of thousands of individual subscribers on all continents and collaborates with an institutional member base including Fortune 500 companies, government organizations, and educational institutions, provide a free research and innovation platform.

Through its open free membership model EAI promotes a new research and innovation culture based on collaboration, connectivity and recognition of excellence by community.

More information about this series at http://www.springer.com/series/15427

G. R. Kanagachidambaresan • R. Maheswar
V. Manikandan • K. Ramakrishnan
Editors

Internet of Things in Smart Technologies for Sustainable Urban Development

Springer

EAI
RESEARCH MEETS INNOVATION

Editors
G. R. Kanagachidambaresan
Associate Professor, Department of CSE
Vel Tech Rangarajan Dr Sagunthala R&D
Institute of Science and Technology
Chennai, India

R. Maheswar
Assistant Dean – Research
VIT Bhopal University
Bhopal, Madhya Pradesh, India

V. Manikandan
Professor, Department of EEE
Coimbatore Institute of Technology
Coimbatore, Tamil Nadu, India

K. Ramakrishnan
Associate Professor, Department of EEE
Pondicherry Engineering College
Puducherry, India

ISSN 2522-8595 ISSN 2522-8609 (electronic)
EAI/Springer Innovations in Communication and Computing
ISBN 978-3-030-34327-9 ISBN 978-3-030-34328-6 (eBook)
https://doi.org/10.1007/978-3-030-34328-6

This Springer imprint is published by the registered company Springer Nature Switzerland AG
The registered company address is: Gewerbestrasse 11, 6330 Cham, Switzerland

To Almighty, Parents, and to our dear students and scholars.

Preface

Internet of Things has got its position even in nanoscale, and it is inevitable for constructing a Smart city. Providing sophistication and amicable lifestyle to individuals is the main motto of the Smart city. The Smart city development mainly depends on components like sensor data, processing capability, transmission ability, and style of the prognosis. Sensor data predictions on the near future are helping us a lot on decision-making. The IoT in a Smart city faces lot of challenges in each stage. (a) The sensing system is exposed to complicated environment, where it undergoes dynamic change in environment and gathering data becomes more complicated. (b) The data from multiple sensors gathering in processing unit creates huge data repositories, making tough time for the data scientists. (c) The ability to transmit data from distance with low energy is a major challenge for researchers working in the sustainable Smart city development. Better predictions help them to face water scarcity, power distributions, and health monitoring of individuals accurately. Industries in Smart city also move towards Industry 4.0 for better productions and predictions. The industries enabled with IoT technology are more aware of market fluctuations and supply chain. The ability of IoT to interact with PLC, embedded, and other physical devices has made sustainable development easier to the researchers. Present systems like PLC boards and embedded systems are designed with IoT add-ons for data gathering and actuations. Automation in industries and Smart city application through IoT technology has improved the knowledge of workers, employers, and serves to be a reasonable solution for labor scarcity. This book addresses the role of Internet of Things in smart technologies for sustainable urban development. The role of IoT in transportation technology in India and different energy harvesting schemes involved for small embedded nodes were discussed. Energy harvesting and managing available energy for sensing the event, the technology involved in oil fields, and power production are also briefly discussed in this book. This book will serve as a solution for unique problems faced in developing Smart city applications.

Chennai, India G. R. Kanagachidambaresan
Bhopal, India R. Maheswar
Coimbatore, India V. Manikandan
Puducherry, India K. Ramakrishnan

Acknowledgment

Our sincere thanks to Editor in Chief and Eliska for giving this opportunity in *Innovations in Communications and Computing*, Springer. We would like to thank the authors and reviewers for their contribution. The support from Vel Tech Rangarajan Dr Sagunthala R&D Institute of Science and Technology, VIT Bhopal University, Coimbatore Institute of Technology, and Pondicherry Engineering College has made this book project possible. I hope this book would be a key for future technology in Smart city development.

Contents

About the Editors

G. R. Kanagachidambaresan received his BE degree in Electrical and Electronics Engineering from Anna University in 2010 and ME Pervasive Computing Technologies from Anna University in 2012. He has completed his PhD in Anna University, Chennai, in 2017. He is currently an Associate Professor, Department of CSE, Vel Tech Rangarajan Dr Sagunthala R&D Institute of Science and Technology. His main research interest includes Body Sensor Network and Fault tolerant Wireless Sensor Network. He has published several reputed articles and undertaken several consultancy activities for leading MNC companies. He has also guest edited several research projects from ISRO, DBT, Government agencies, special issue articles and serves as an editorial review board member for peer-reviewed journals.

R. Maheswar has completed his BE (ECE) from Madras University in the year 1999, ME (Applied Electronics) from Bharathiar University in the year 2002, and PhD in the field of Wireless Sensor Network from Anna University in the year 2012. He has about 17 years of teaching experience at various levels and presently working as an Associate Professor in the School of EEE, VIT Bhopal University, Bhopal, India. He has published 50 papers at International Journals and International Conferences. His research interest includes Wireless Sensor Network, IoT, Queueing theory, and Performance Evaluation.

V. Manikandan is serving as a Professor in the department of EEE, Coimbatore Institute of Technology, Coimbatore, India, for more than two decades. He has guided eight doctoral scholars in the last one decade, and three of them have contributed significantly in control and communication areas. The research outcomes over these years have been published in refereed international journals. Some of the research works have been presented in leading international conferences in India and abroad. He has obtained many research grants in the last 10 years from various Indian funding agencies like AICTE, UGC, DST, MHRD, etc. for executing various research projects, conducting faculty development programs, and infrastructure developmental activities in the Institute campus and in the neighboring ecosystem. He has consultancy experience on energy auditing and energy optimization. His areas of interest include modelling and synthesis of efficient lossless power electronic conversion circuits to cater power interface between PV/wind sources and grid, development of soft computing-based control strategies for smart power systems, and embedded electronics.

K. Ramakrishnan has completed his BE (EEE) from Anna University in the year 1993, ME (Control System Engineering) from Bharathiar University in the year 1995, and PhD under the title, *Robust Stability Stabilization of Dynamical State Delayed Systems* from IIT, Kharagpur, West Bengal, in the year 2014. He has about 23 years of teaching experience at various levels and presently working as an Associate Professor in the Electrical and Electronics Engineering Department, Pondicherry Engineering College, Pondicherry. He has published 35 papers at International Journals and International Conferences. His research interest includes IoT and Control Theory and Applications.

IoT in the Field of the Future Digital Oil Fields and Smart Wells

Ghazanfar Latif, Jaafar M. Alghazo, R. Maheswar, A. Sampathkumar, and S. Sountharrajan

1 Introduction

The IoT has gained its popularity in the global information industry and petroleum explorations industry. The IoT connects everything with the Internet because it is an intelligent network; this helps the devices to communicate and exchange information. What IoT achieves is intelligently, locating, managing, identifying, monitoring, and tracking various things. It is considered an expansion and extension of the Internet-based network, which is the communication expansion from things to things and human to human. Objects that surround people are going to be connected into networks in one way or another in the IoT paradigm. The smart technologies, including, sensor technology, radio frequency identification (RFID) will be embedded in many applications. The evolutions of technology such as storage, battery capacities, and computing power are becoming smaller in size and lower in cost. Hence, this helps developing small electronic devices with capabilities like computing, identification, and communication that can be embedded in other systems, facilities, and devices [1].

G. Latif (✉) · J. M. Alghazo
Department of Computer Science, Prince Mohammad Bin Fahd University, Al-Khobar, Kingdom of Saudi Arabia
e-mail: glatif@pmu.edu.sa; jghazo@pmu.edu.sa

R. Maheswar
Assistant Dean – Research, VIT Bhopal University, Bhopal, Madhya Pradesh, India

A. Sampathkumar · S. Sountharrajan
School of Computing Science and Engineering, VIT Bhopal University, Bhopal, India
e-mail: s.sountharrajan@vitbhopal.ac.in

G. R. Kanagachidambaresan et al. (eds.), *Internet of Things in Smart Technologies for Sustainable Urban Development*, EAI/Springer Innovations in Communication and Computing, https://doi.org/10.1007/978-3-030-34328-6_1

The wireless sensor network (WSN) technology can be utilized in smart oil wells [2]. WSN is a group of sensors for recording and monitoring the condition of a specific environment. Some WSN applications are monitoring pipelines, equipment condition, and natural gas leaks. The system monitors three sectors of the industry and displays reports of the condition of the well. They sense pressure, vibration, pipelines defects, and temperature. The control center receives all of these important parameters and plans for all the steps that need to be taken. Oil theft is prevented because everything is recorded and stored by WSN. The life of the well is extended. The system monitors the holes' pressure that is created to elicit oil and gas from the earth. The high pressure causes the oil to leak, so the system makes sure to avoid that. In addition, the system decides the number of fluids that are needed to be injected inside of the well to displace oil. Furthermore, the system enhances the oil flow through the hoses by mixing it with carbon dioxide and reduces the thickness of the oil. The drawbacks of the system are the wires and heavy tools that cost a high budget and space. The intended solution is a wireless monitoring system and collecting the data remotely and later transmitting to cloud servers to analysis.

2 Literature Review

The Petroleum Analytics Learning Machine (PALM) is a learning-based machine that has a system to manage the gas and oil fields using IoT [3]. PALM has machined analytics applications, which are big-data-centric. It uses computational machine learning as well as predictive and prescriptive analysis techniques to reduce the losses and increase production. The complex machine has smart sensors that bring IoT together to make it a digital oil field. The machine combines artificial intelligence with machine learning to deliver better performance and lower the cost. PALM's software works on utilizing the support vector based machines, which excel the pipeline system performance. In the application, the algorithm functions by open source codes programmed and combined together as tools. The tools hold all of the geological and geophysical attributes that measure the volume of the gas and oil. The data gathering is done by the IoT communication network. The PALM system integration database stores the collected data. Data is processed and cleaned from noise then normalized to be ready for distribution between equipment and tools. The system utilizes codes and boosts the machine neural network. PALM's estimations of the volume and depth of wells are accurate, with low error rate. The distributed control aides in calculations and helps scientists in integrating geology with IoT. The model is still not at its final version as it is constantly developing. The drawback is the integration system, which is not fully successful yet. It is not easy to create a fully distributed network system, which is 100% reliable. The solution to overcome these issues is by designing better sensors that minimize the tasks as much as possible.

Magnetic flux leakage (MFL) is a method to detect corrosion, pitting, and wall loss in steel structures. MFL system works by sensors on the pipelines and detects the different defect types [4]. The targeted pipelines are covered with sensors to measure and collect data. The machine-learning algorithm calculates the parameters that are needed to optimize the performance of the neural network. The financial losses of the pipeline defects are huge, and it will be decreased by using IoT. The intelligent tools locate the magnetic flux leakage with the help of the signals and ultrasonic waves to the pipeline defect. Every defect has a pattern of behavior, and the solution can be applied by studying those behaviors. When the sensor has high axial amplitude, it explains the damage and its size. Several solutions have been suggested, but the most effective one is combining the MFL signals along with artificial neural networks to achieve the maximum reliability and estimate correct results to solve the problem. There is a problem which is the relationship between the depth of the amplitude generated by the signals which cannot always be analytically described. Complexity is an obstacle and identifying the relationship to detect the defects is not easy to implement. The solution could apply a dynamic neural network.

Oil wells need a monitoring system that tracks all the needed parameters [5]. The system consists of layers, and each layer deals with a required task. It has the data acquisition and monitoring layer, then the transmission layer, and finally the data analysis and production management layer. Collecting and processing the data in the first layer is accomplished by the sensors and actuators. They test the power, query locations, and detect temperature. The transmission layer transfers all the data that has been collected by the acquisition layer and managed in a secure channel. 3G, GSM, and ZigBee are some of the technologies, which are used in transferring data. The final layer is the layer that combines the collected and processed data and displays the business model that consists of flowcharts and graphs. It is all done by the network wire protocol. The used algorithm called six Sigma, and its major function is processing the big data. The main issue of the system is the low pressure of the well oil. The system sometimes does not alert the control center, and this will inhibit the daily oil production from increasing. The solution is by inventing an algorithm that calculates the fluids that need to be injected into the pipeline and display the parameters; this algorithm will allow the pressure to go back to normal. Furthermore, the algorithm detects the abnormality of the fluids. Any malfunction should be analyzed and the changes of the parameter should be automatically done.

Big data analytics in petroleum engineering is important because it includes all unstructured and multi-structured data [6]. The collected data are used to develop images of subsurface layers. Extracting gas and oil using drills and exploration became much easier by using IoT technology. The rig transforms into a digital one where the operators finish the job from a distant. The main method is using big data to control velocity, variety, and value. Big data has been profitably used in automated drilling and detecting services. Big data has enhanced the design of smart digital oil fields [7]. Drills and pipeline platforms changed drastically, and the extraction cost reduced. The pattern-based analysis is more desirable to scientists because of the huge impact it has on IoT. Reducing hydraulics and heavy machines is a big goal, which is more likely to be achieved in the next few years. Some information is still lacking, but the researches are still on the path to find solutions and more data processing to reach the maximum accuracy. Data infrastructure in big data analytics aided in the discovery of oil fields. Compromising stacks to get values out of the data. All the processed data include CAD and daily drilling reports. The model faces some technical challenges, such as I/O on the network, fetching shared data effectively, and having multi-label connections. Years of research and progression will solve most of these problems and build an integrated smart oil well, which is automated and digital.

Enhanced oil recovery (EOR) is the recovery of non-extracted oil by natural or enhanced pressure to make it easier to extract. There are three techniques for EOR, which are chemical injection, gas injection, and thermal. One of the old methods that have been used in oil recovery is the thermal method [8]. As expected at this point of evolution of technology, the thermal method is somewhat not practical. Engineers came up with a new method that is much more economical to maintain daily oil production at its best peak. The EOR system deploys advanced IOR technologies that control the field. The major tasks that the system performs are heat injections, solvent injections, and surfactants injections. The main three methods help in recovering oil from rocks. The process of injecting the right amount is complex and has various stages. The findings were quite remarkable. The daily oil barrel production number has increased by 10%. One of the advantages of the method is applying it in the early stage of the life cycle of the reservoir. The technology is still new and still in the development stage. It encountered many issues such as model complexity. Another issue is how harmful it could be to the environment. These issues need to be solved and finding a better correction to the system before commercial use. A large amount of data is processed and analyzed over a high-speed network. The stored large data have different formats as well as different vendors. Sometimes data cannot be traced back, so any carried information might be considered lost. The proposed system model is capable of obtaining the lost data and generating the process of oil extraction. Integrating the mapping of annotations of any dataset is a significant solution. The work focuses now on using domains to integrate seamless data [9].

It has not been a loss to the oil industry when Machine to Machine (M2M), Big Data, and Internet of Things technologies were developed in other areas. It rather has created oil and gas new invention theme, which is known as the digital oil field. Based on Group of Cleantech definitions, the category of digital oil field involves services, related business models, and technologies concentrated in upstream oil and gas activities on the processes and tools for information and data management [10]. Numerous leaders have already worked on their digital oil field initiatives in oil and gas companies' integration; for instance, Shell's Smart Fields, Chevron's Fields, and BP's Field of the Future. Moreover, leaders will carry on looking for new technologies for the sake of decreasing operating costs, as well as increasing output of productions. The term digital oilfield comprises various elements of technology all over the software and hardware solutions, which are circulated everywhere the whole upstream oil and gas activities suite. They are all considered as attempts to provide analysis, transmission, subsequent automated action, and data capture in real time, hence, optimizing performance of production. Nonetheless, the oil and gas section utilize IoT and Big Data concepts to its activities of upstream. Data analysis, communications technologies and networking, and hardware innovations are qualified as new processes of E&P known as the digital oil field.

Smart oil fields have been deployed and proposed for oilfields offshore remotely. Sensors from a wide range included in a smart oil field such as density of gas, temperature, and pressure pipeline, which produce collective data daily on the terabytes' range, require real-time analysis. Solutions of smart oilfield that exist use remote data centers over the cloud and satellite communication, which are not feasible because of significant delay of transfer. A solid allocation model of resource using computing edge is presented as a solution to consider computational capacity with limited connectivity and application intensive resource in the oilfields. The model that is proposed assigns tasks efficiently to a suitable cloud or edge resource to address the constraint of real-time applications in order to obtain robustness. Evaluation of the results displays that the proposed model can upgrade the system performance significantly compared to architectures of conventional based on cloud [11].

There is a limitation to the new oil and gas reserves in recent years, as it is generally spread in environments that are harsh, such as the deep sea, the desert, and the Arctic environment. The activities of production and exploration have an influence immeasurably on the environment. The pressing challenge of the oil companies is, knowing the ways of rationally implementing the existing technology so that the recovery of oil gets improved. The smart oilfield is a method of a loop that is closed asset management, which can achieve data acquisition in real time, monitoring in present, the implementation of the optimization and decision-making, interpretation in real time, oilfield, oil well, and other assets can be connected to one another to assort the management and operation, and thus the effective direction and way to enhance recovery of oil. Recently, with the mature and progress of the monitoring technology of the dynamic reservoir, managing horizontal wells, recovery of oil prospects is improved by the smart oil field. Intelligent oilfield is a complicated system which consists of a process, technology, and people organic combination. Raising the values of the assets through managing reserves of oil and gas in real time and on demand is the final objective. This can be done by reporting and collecting a massive data. Furthermore, under an environment of decision collaboration, timely informed decisions are required [12]. Table 1 shows the different methods used in digital oil fields and the accuracy.

Table 1 Total worldwide oil resources

Resources	% Share of resources
Oil sands bitumen	30
Conventional oil	30
Extra heavy oil	25
Medium heavy oil	15

2.1 Statistics of Oil Reserves, Production, and Consumption

It is difficult to relate the resources of oil in all the worldwide; this is because there are several areas around the world which are still not fully known geologically and unexplored by drilling. As shown in Table 1, oil sands bitumen represent 30% as well as conventional oil. However, the extra heavy oil represents 25%, while medium heavy oil only 15% [13].

The universal competition for producing oil has increased in the last decade. Many industrializing countries entered this field, including India and China. However, some nations are still in the top globally such as the USA, which represents 18% of worldwide oil production. Followed by Saudi Arabia 12%, Russia produces 11%. Canada and China account for 5% of global oil production. Table 2 shows the top 5 oil producing countries [14].

According to the energy information administration (EIA), the total amount of crude oil consumption worldwide is about 93 million barrels per day. [15]. The USA is the largest consumer of crude oil with 20% of world consumption, which is equivalent to 19.69 million barrels per day. China is the second consumer country after the USA, and it represents 13% of worldwide oil consumption, with 12.79 million barrels per day. After that, India and Japan have convergent values, 5% and 4.44 million barrels per day for India, yet 4% and 4.01 million barrels for Japan. Russia represents also 4%, but with 3.63 million barrels. Table 3 shows the top 5 oil consumption countries.

Table 2 Top 5 oil producing countries

Country	Million barrels per day	% Share of world's total production
USA	17.87	18
Saudi Arabia	12.42	12
Russia	11.40	11
Canada	5.27	5
China	4.82	5

Table 3 Top 5 oil consuming countries

Country	Million barrels per day	% Share of world's total production
USA	19.69	20
China	12.79	13
India	4.44	5
Japan	4.01	4
Russia	3.63	4

3 Current Methods for Oil Fields Management and Drilling

Large companies leverage the usage of intelligent processes and technologies in drilling oil fields to achieve their strategic goals. The first method to manage the smart oil fields is creating intelligent field centers to make decisions in a collective way. To ensure the success of these centers, a project methodology should be followed. The methodology should start with ensuring a clear understanding of the intended plans and it concludes with maintaining continuous improvement. The methodology has five stages, which are assessment, design, construction, implementation, and continuous improvement as shown in Fig. 1. In the assessment stage, the teams that will be working on the project and the stakeholders will determine the requirements. In this stage, the activities and workflows including the need for software, hardware, data, and visualization will be defined. In the design stage, there can be an improvement. The construction stage includes testing and installing the hardware and software as well as the other infrastructure that will be needed in delivering the reservoir performance and technical planning workflows. The stage embraces the civil and mechanical task aspect of the environment, such as electricity and lighting. The defined work plan, associated software, and infrastructure will be customized in the implementation stage. This will need defining the technical aspects such as software integration, process implementation, and process optimization as well as the human aspects such as cross-domain integration and training. Finally, the continuous improvement stage will include regular meetings to follow up with the workflow. Those meetings will help the reservoir management engineers to understand the problems and provide the best solutions and recommendations to solve them [16].

The management system of oil industry shown in Fig. 2 consists of some fundamentals and elements divided into two independent components. The fundamentals focus on the most important management principles to have an efficient and

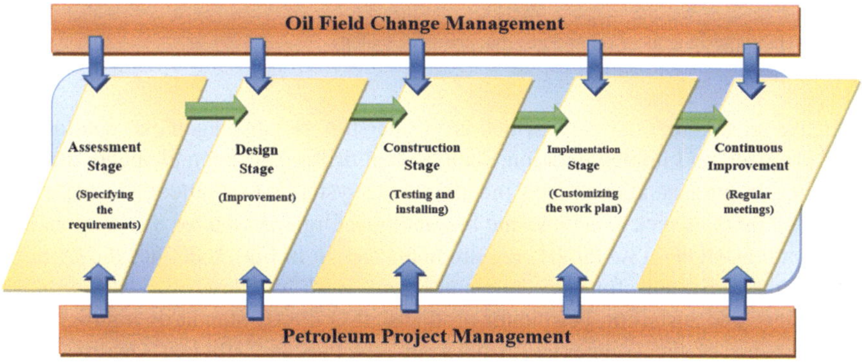

Fig. 1 The stages of intelligent field centers

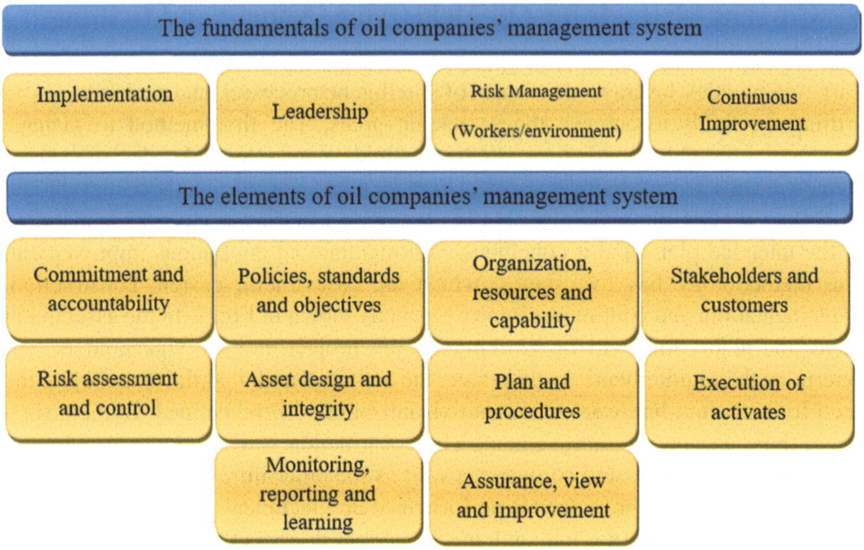

Fig. 2 The fundamentals and elements of management systems

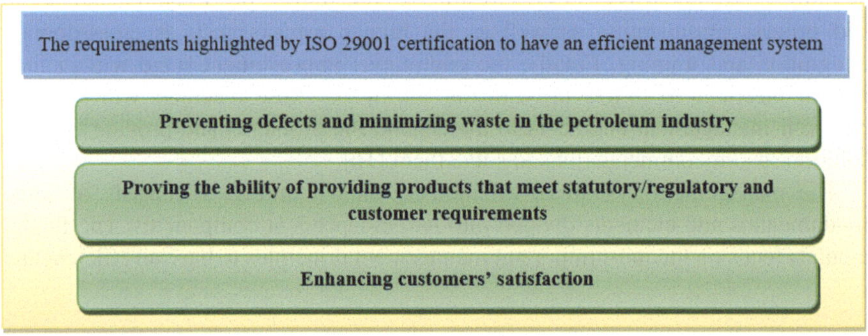

Fig. 3 The requirements of oil and gas management system

effective management system, which are leadership, risk management, continuous improvement, and implementation, such as well drilling. The ten elements include a set of exceptions and overviews that define the system's best outcomes [17].

Oil companies seek to achieve the International Organization for Standardization certifications (IOS). ISO 29001, requirements shown in Fig. 3, defines the Quality Management System for the companies that work on designing, developing, producing, and installing petroleum products. ISO 29001 standards highlight the requirements of the companies to have an effective and efficient Quality Management System. The first requirement is proving the ability to provide products that meet statutory/regulatory and customer requirements. The second requirement is enhancing customers' satisfaction through many improvement processes and providing requirements that meet their needs. The third requirement focuses on preventing defects and reducing waste. Achieving the certification assists in showing the commitment to industry best practices and managing operations effectively [18].

4 Challenges in Oil field Management Systems

The IoT is still in its early stages, which is because of the many challenges that hinder the future growth. The main obstacles include the lack of skilled workers, lack of theory, and lack of systematic research for reaping the benefits [19]. Distributed intelligence designs are limited and there are many problems that need to be solved regarding the sensors, cameras, power nodes, and identification. One of the main challenges is the architecture [1]. Data integration in different environments is tough due to changes in the earth's layers or between devices. The system has to combine different sources in order to determine which feature to use, how to extract the oil, what drill to use, and how to determine the depth. The technology has to be fully functional for every environment and for all cases. RFID technologies differ as well as wireless networks. Different requirement for each communication can take time and be costly.

Sometimes the unnecessary dependencies are going to blackout the migration of connecting as many things together. Introducing a successful platform requires successful hardware and here lays the challenge. The systems have to be in a minimum size and lightweight with a wireless identifiable system. Some levels diverge because they go from kbps to mbps. It is a challenge to have an ultra-cost with low-powered hardware because when the system is on sleep mode, it will not be able to receive the RF signals and it will cause an economic loss. The system cannot combine all the features together; low cost, low power, and size with integrated data could be a challenge as well [19].

Another challenge is privacy and security. It has to exceed traditional security methods because the system deals with oil. The challenge is designing a secure system, which is capable of doing all the tasks as well as maintaining a high level of security. In addition, the system has to analyze the oil parameters and transfer the data between the platform and the terminal through the connection protocol. It could be a challenge to minimize and achieve the desired level of accuracy because all of the objectives are still in the developing phase. The challenge could be in the routing process, which is selecting the best path for data movement. The best path can be chosen based on the number of hops, bandwidth, and costs. Another problem is big data, which is massive data, whether unstructured or structured. The dataset could be big data when it meets the four V's, which are volume, variety, velocity, and veracity. It is hard to deal with it by using traditional software techniques and database methods. An example of big data could be a huge amount of data collected from deploying sensors [20].

5 Improving the Functioning of the Oil and Gas Industry with IoT

The oil and gas industry face many obstacles, which should be eliminated in order to enhance oil and gas production and getting higher revenues. In the meantime, the IoT started to enhance the operation of the oil and gas sector globally. The Internet of things has a big influence; this is because it helps to get rid of many problems the oil and gas industry face. According to [21], there are many techniques that will help to revolutionize the oil and gas industry, which includes the preventative maintenance for remote oil and gas equipment. This will help the large oil companies to monitor the wells more easily. For example, a company that monitors more than 50,000 wells will spend the big amount of money on the equipment. Furthermore, a single malfunction in any of the machines will cause slow production. The IoT technologies help in equipment maintenance and protect the equipment. This is because the IoT device can be deployed to remotely monitor equipment. This will limit losses as well as increase the production. Also, improved operational efficiency, which means the big data analysis and remote visibility will support the companies to manage their assets and use their findings to optimize production. In addition, real-time data, which will help to capture the oil production in real time through the embedded sensors will help the companies to collect the needed information related to the oil fields from anywhere in the world. The wide adoption of IoT could raise global GDP. Moreover, fewer safety risks, the Internet of things could minimize the risk that could show up as a result of potential issues in the oil industry field. The real-time data, will help to capture the oil production in real time by the embedded sensors. Figure 4 shows the proposed organization of smart wells.

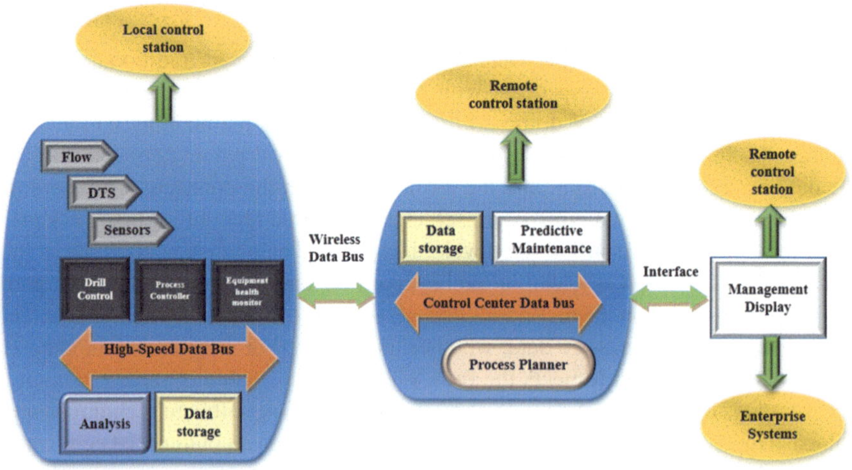

Fig. 4 Proposed organization of smart oil wells

Since the appearance of Internet of Things (IoT), many industries have been practicing and applying its technologies and concepts. Industries have also been driving IoT security efforts and standardization. A new concept on Industrial Internet has evolved from the integration of IoT in many industries. The Industrial Internet Reference Architecture (IIRA) has been recently published with standards for the integration of IoT in industries. Industries like the oil and gas industry can utilize the full capability of the IoT within the processes of upstream and downstream. There is a growth in terms of system integrators and device suppliers' ecosystem. There are protocols, protocols of communication more specifically, interfacing one to another for the sake of integrating seamlessly like Data Distribution Service for Foundation of OPC. In addition, there is a chance to implement the maintenance and improve the safety to have better production, as well as enhance operations by extracting helpful guidance from the developed case studies. A severe cut of cost and a widespread time investment are required for these initiatives, which can be a key to automating tasks of intensive labor, dilating operation of remote, and lowering cost of production, which all participate in the bottom line. The main challenge is security however, this issue hasn't been ignored and the appropriate methods do exist. Data residency is another big challenge because it is a technical issue but at the same time legal and political, many regions will find it too difficult to remove.

Oil and gas organizations are considered very complex. Volumes of data with business are exponentially expanding but bounded by the data quality, IT systems traditions, and ability of staff. As the technology of big data has been developing lately, the big data capabilities benefit oil and gas organizations to perform analysis on data under large volumes, transforming reactive to proactive in decision-making, and optimizing the whole development, production, and exploration phases. The forum of world economy in 2011 suggested that the new economic oil is Big Data. Furthermore, technology on big data in gas and oil industry is an ongoing stage of experiment. The sizable data application is beneficial in terms of breaking through the leaders of oil and gas companies' bottleneck development which almost all leaders of the oil companies believe. Moreover, they recognized the value technology of big data on restructuring the industry. The big data is a set of complex and huge data, which is a field that analyzes information and deals with large and complex data. This will help to deal with the software applications of processing traditional data. Thousands of sensors are used by oil and gas companies that are installed in facilities' surface and wells subsurface in order to provide continuous conditions of environment, collecting of data, and assets real-time monitoring. In development and exploration, professionals execute operational and strategic tasks in making decisions with the support of analysis of large data. For instance, in terms of data of real time, new insights can be delivered by the company that can enhance efforts of exploration, which is considered a big help for operating teams. Based on real time and historical data in terms of production, future performance can be predicted by big data. Based on the collected data, equipment of maintenance, big data can support to restrain down time and to structure a model for predictive maintenance in oil and gas companies. Figure 5 shows the oil and gas industry big data conceptual architecture.

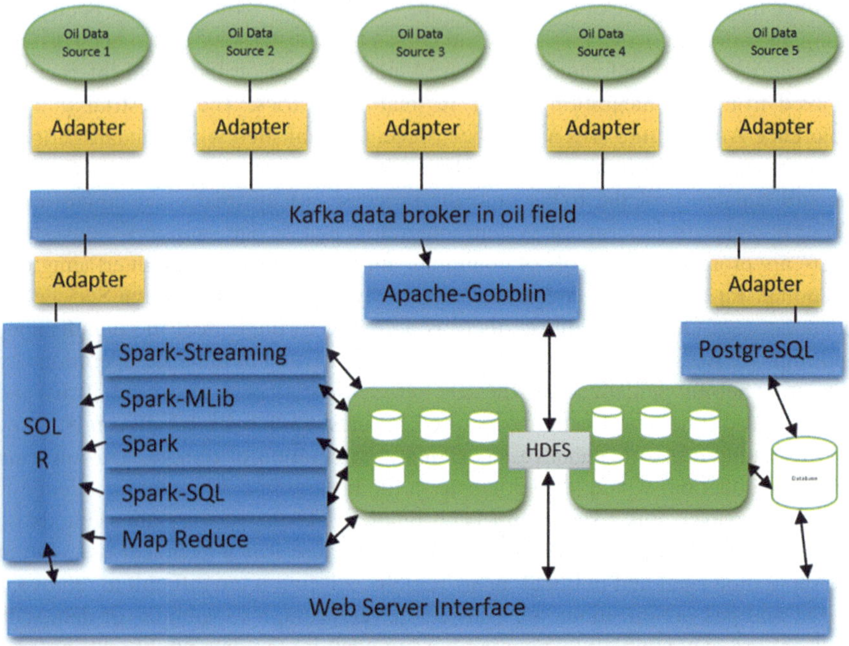

Fig. 5 Oil and gas industry big data architecture

A system for improving the drilling process could be a good example of using IoT in drilling oil fields. The first part of the system maintains the used equipment and speeds up the process of requesting new devices. It monitors the level of oil. When an error occurs, the system will be turned off automatically. Those errors will be detected by sensors, which collect data from the field, such as oil value, oil low value, oil very low value, oil high value, oil very high value, and oil set point. In the system in Fig. 6, LT refers to the level transmitter while LC refers to the level controller; PC refers to the pressure controller while PCV refers to the pressure control valve. Finally, XV is the shutdown value. The second part is the thin client application. This part provides many services such as determining the initial values, viewing the values of gas and oil, the timer of reading the sensor, and a web application that enables the users from monitoring the system from any place. The web application has various types of users, such as engineer, operator, and manager. The administrator will assign the privileges for users based on their responsibilities.

Development and economy of cities dependent on oil industry are confronting at present extreme social, environmental, and economic issues. These cities industrial structure has a comparatively traditional pattern of industry, which are based on petroleum and intensive petroleum economy with relatively low efficiency and technology levels. Metropolises in China have stepped inside smart cities in the widest range, and cities that are based on petroleum are still lagging behind.

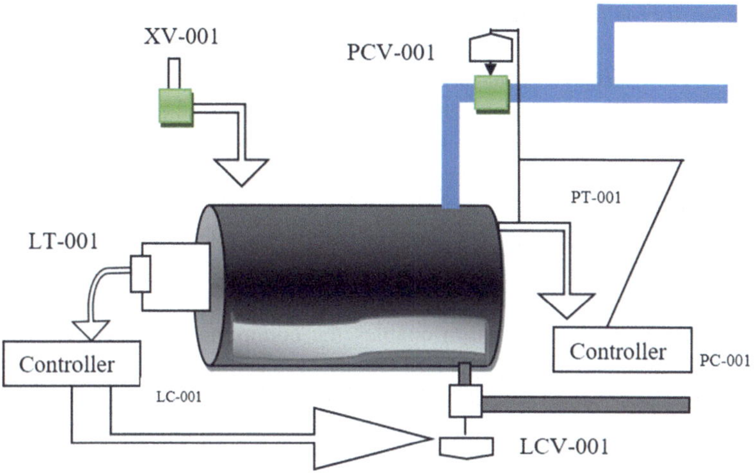

Fig. 6 Proposed IoT Sensors based system state for drilling oil fields

Hence, metropolises in China have good environmental and economic potentials, especially in the energy field, yet, with stage of depletion metropolises must show conditions of living favorably. Cities based on petroleum evolving into smart cities under the development of smart city trend are inevitable. However, oilfield-based oil exploitation as the fundamental service of cities based on petroleum drives economic development. Development of smart oil field is therefore the transference of cities based on petroleum key point. The idea of digital oil field is where the smart oil field concept evolves. The digital oilfield core is networking and digitalization that underlines collection of data. Smart oilfield presents human and artificial intelligence by contrast, hence, underlining their integration and focusing on sharing knowledge, managing intelligence, making decisions scientifically, and mining data. A developed smart oil field future reference is provided. Harmonious, green, efficient, and intelligent are the conclusion of the development of smart oilfield targets [22].

6 Benefits of IoT in the Oil Field

There are many benefits of IoT including: comprehensive perception, obtaining data and information from any spot by using sensors, two-dimensional barcode, and RFID. This could make the communication systems and information be invisibly embedded in the environment. Workers will be able to interact with the machines using sensors network. Location and objects identifications are considered as identification technologies. To implement overall perception, it is required to have recognition and identification of the physical world. Using sensors such as gas pressure sensors and gas pressure switches caused a notable improvement in the safety and security of workforce, assets, and operations.

Reliable Transmission: Objects' information can be available at any time through different types of available telecommunication networks, the internet, and radio networks. Switching technologies, gateway technologies, networking technologies, and wired and wireless transmission technologies are types of communication technology. In addition, IoT creates machine-to-machine (M2M) interaction, which is considered as the main implementation technology of the Network of Things. M2M innovations have been used in the petroleum industry to maintain the health of workers from life-threatening activities and to increase the production of gas and oil. Furthermore, M2M interaction enables the communications and connections between M2M, human to machine, and mobile to machine.

Intelligent Processing: Several technologies that deal with intelligent computing such as cloud computing, which supports IoT data applications by gathering IoT data and placing them in a form of databases. The cloud technologies allow the workers to monitor the system effectively to prevent accidents such as gas leakage. The connected systems can be programmed to send an alert to all the devices in the system when it begins to fail.

Cost Reduction: More than $100 per barrel of longstanding oil price has led to even higher increase in oil prices throughout the years. According to Hamilton, supply in fresh sources opened up by technologies of new extraction, which suggests $20–$30 less per barrel as equilibrium of a new price. Lowering oil prices is beneficial in terms of exposing the inefficiency of oil and gas companies as well as leading the efficient companies to find their ways of preserving the bottom and top lines. The new suite of technologies engages in giving companies a hand to deal with these challenges, which is good for both gas and oil industries. IoT has been simmering for a while, which basically integrates analytics, communications, and sensing capabilities. At any point it will boil over, which means the improvement of the core enabling technologies are expected to reach a point where the adoption of widespread of core enabling technologies seems more likely. The promise of IoT does not include the assistance of Oil and Gas companies in managing their customer relationships supply chains or existing assets. As a matter of fact, a new asset of information businesses concerning these elements was entirely created by IoT technology. Therefore, there is no wonder that the one-size-fits-all IoT solution came up. Operations optimization, new value creation, and reliability improvement are the three business objectives the oil and gas industry has related to the deployment of IoT. In one of these three objectives, every oil and gas segment could be able to discover the benefit of the IoT initial efforts and new information sources, which can enable them. Production and exploration are examples of upstream organizations, which concentrate on optimization. It examines various groups of cross disciplinary, physics, and non-physics data. Storage, pipelines, and transportation are examples of midstream organizations that concentrate on opportunities of new commercial and integrity of higher network, which aims to designing an infrastructure of enabled data to discover certain benefits.

7 Conclusion

The IoT has progressed in the previous 10 years; this includes the huge improvement in oil field industry, which resulted in changing the gas and oil industry remarkably. The tasks became automated, faster, and economically much more conserved, which helped to increase the daily production. IoT is changing the oil industry and extraction techniques and turning it into a whole new level of technology evolution. The integrated network and machine learning models in smart wells overcame all the obstacles and struggles that scientists had been facing. In the near future, it is possible that robots are going to replace the operators in the control center of rigs and oil fields. This will help to improve the overall functions that relate to oil field. The new intelligent systems are going to be capable of doing all the necessary functions to extract oil, detect the right amount needed to inject and measure all the volumes, and generate full reports of the results.

References

1. Bandyopadhyay, D., & Sen, J. (2011). Internet of things: Applications and challenges in technology and standardization. *Wireless Personal Communications, 58*(1), 49–69.
2. Mohamed, A., Hamdi, M. S., & Tahar, S. (2015, August). A machine learning approach for big data in oil and gas pipelines. In *2015 3rd International Conference on Future Internet of Things and Cloud* (pp. 585–590). IEEE.
3. Carvajal, G., Maucec, M., & Cullick, S. (2017). *Intelligent digital oil and gas fields: concepts, collaboration, and right-time decisions*. Oxford: Gulf Professional Publishing.
4. Anderson, R. N. (2017, December). 'Petroleum Analytics Learning Machine' for optimizing the Internet of Things of today's digital oil field-to-refinery petroleum system. In *2017 IEEE International Conference on Big Data (Big Data)* (pp. 4542–4545). IEEE.
5. Aalsalem, M. Y., Khan, W. Z., Gharibi, W., & Armi, N. (2017, October). An intelligent oil and gas well monitoring system based on Internet of Things. In *2017 International Conference on Radar, Antenna, Microwave, Electronics, and Telecommunications (ICRAMET)* (pp. 124–127). IEEE.
6. Xu, B., Wang, W., Wu, Y., Shi, Y., & Lu, C. (2017, March). Internet of things and big data analytics for smart oil field malfunction diagnosis. In *2017 IEEE 2nd International Conference on Big Data Analysis (ICBDA)* (pp. 178–181). IEEE.
7. Mohammadpoor, M., & Torabi, F. (2018). Big data analytics in oil and gas industry: An emerging trend. *Petroleum* (In Press). https://doi.org/10.1016/j.petlm.2018.11.001.
8. Kokal, S., & Al-Kaabi, A. (2010). *Enhanced oil recovery: challenges & opportunities*. World Petroleum Council: Official Publication, 64.
9. Chelmis, C., Zhao, J., Sorathia, V. S., Agarwal, S., & Prasanna, V. (2012, January). Semiautomatic, semantic assistance to manual curation of data in smart oil fields. In *SPE Western Regional Meeting*. Society of Petroleum Engineers.
10. Reece, C. A., Hoefner, M. L., Seetharam, R. V., & Killian, K. E. (2008, January). An enterprise-wide approach to implementing 'digital oilfield'. In *Intelligent Energy Conference and Exhibition*. Society of Petroleum Engineers.
11. Hussain, R. F., Salehi, M. A., Kovalenko, A., Salehi, S., & Semiari, O. (2018). Robust resource allocation using edge computing for smart oil fields. In *Proceedings of the International Conference on Parallel and Distributed Processing Techniques and Applications (PDPTA)*, Las Vegas, pp. 204–210.
12. Hao, X. (2013). Intelligent oilfield work processes-turning data into effective decision and actions. Master's thesis, University of Stavanger, Norway.

13. Hendricks, T. A. (1965). Resources of oil, gas and natural-gas liquids in the United States and the world. *United States Department of Interior, Geological Survey Circular, 522*.
14. Dale, S., & Fattouh, B. (2018). Peak oil demand and long-run oil prices. *Energy Insight*, p. 25.
15. U.S. Energy Information Administration. Retrieved from https://www.eia.gov/tools/faqs/faq.php?id=709&t=6
16. Barghouty, M. F., Al-Dhubaib, T. A., Jama, A. A., & Jaimes, O. (2010, January). Intelligent field centers (IFCs): Integrating people, processes and technologies to optimally manage giant fields. In *SPE Intelligent Energy Conference and Exhibition*. Society of Petroleum Engineers.
17. Qu, T., Thürer, M., Wang, J., Wang, Z., Fu, H., Li, C., & Huang, G. Q. (2017). System dynamics analysis for an Internet-of-Things-enabled production logistics system. *International Journal of Production Research, 55*(9), 2622–2649.
18. The International Association of Oil and gas Producers. (n.d.). *Management systems*. Retrieved July 25, 2019, from https://www.iogp.org/oil-and-gas-safety/managementsystems/
19. Gilchrist, A. (2016). *Industry 4.0: the industrial internet of things*. New York: Apress.
20. Ali, Z. H., Ali, H. A., & Badawy, M. M. (2015). Internet of Things (IoT): definitions, challenges and recent research directions. *International Journal of Computer Applications, 975*, 8887.
21. Tao, Z. (2010). IOT's Application in the "Digital Oil Field". *Telecommunications Science, 26*(4), 25–32.
22. Liu, H., & Ma, X. (2019). Smart oilfield development and transition of petroleum-based cities. *Frontiers of Engineering Management, 6*, 299–301.

Ghazanfar Latif is research coordinator (Deanship of Graduate Studies and Research) and PhD scholar at University of Malaysia Sarawak, Malaysia. He earned his MS degree in computer science from King Fahd University of Petroleum and Minerals, Saudi Arabia in 2014 and BS degree in computer science from FAST National University of Computer and Emerging Sciences, Pakistan in 2010 by remaining in Dean's honor list. Throughout his educational carrier, he got a number of achievements like full scholarship for FSc, BS-CS, and MS-CS. He worked as an instructor at Prince Mohammad bin Fahd University, Saudi Arabia for 3 years in CS Department and has 2 years industry work experience. His research interests include image processing, artificial intelligence, neural networks, and medical image processing.

Jaafar M. Alghazo obtained his PhD and MSc in computer engineering from Southern Illinois University Carbondale in 2004 and 2000, respectively. He joined Prince Mohammad Bin Fahd University (PMU) as founding Dean of the College of Computer Engineering and Science and held various positions including Dean of Graduate Studies and Research, Dean of Institutional Relations, and Dean of Continuing Education and Community Service. Currently he is an assistant professor at PMU. His research interests include modeling and realization of biological mechanism using CAD and FPGAs, modeling and realization of arithmetic operations using CAD and FPGAs, low power cache design, and assistive technology for students with disabilities.

R. Maheswar has completed his BE (ECE) from Madras University in the year 1999, ME (applied electronics) from Bharathiyar University in the year 2002 and PhD in the field of wireless sensor network from Anna University in the year 2012. He has about 17 years of teaching experience at various levels and is presently working as an associate professor in the School of EEE, VIT Bhopal University, Bhopal. He has published 40 papers at International Journals and International Conferences. His research interest includes wireless sensor network, IoT, queuing theory, and performance evaluation.

A. Sampathkumar received his bachelor's in information technology in 2009, master's in mainframe technology in 2012 and PhD degree in 2019 under Anna University Chennai. He has 8 years of academic experience and is currently working as an assistant professor in the school of CSE, VIT Bhopal University, Bhopal. He had published several articles in peer-reviewed journals and member of CSI societies. His research interest includes artificial intelligence, data mining, machine learning, IoT, data analytics, and optimization techniques.

S. Sountharrajan has completed his PhD from Anna University in information and communication engineering in the year 2017 and is currently working as a senior assistant professor in the School of Computing Science and Engineering at VIT Bhopal University. He is the Program Chair for B.Tech/ M.Tech CSE Specialization in Artificial Intelligence and Machine Learning courses. He has also published 35 papers in International Conferences and journals. His research interest includes artificial intelligence, machine learning, data mining, data analytics, optimization techniques, and IoT.

Reinforcement Learning Concepts Ministering Smart City Applications Using IoT

R. Dhaya, R. Kanthavel, Fahad Algarni, P. Jayarajan, and Amita Mahor

1 Introduction

The word "smart city" is a name granted to a city that uses information and communication technologies (ICT) to improve the standard and quality of city facilities such as energy, transference, and efficacies in order to minimalize reserve ingesting, wastage, and general costs. The aim of the smart city is to upgrade the city that renders core infrastructure and provide a quality of life to its people and a clean and sustainable environment smart solutions. The features of smart city are to improve infrastructure and services which include access to water, electricity, affordable homes, education and health services, and IT connectivity through smart technology. The need of smart city is to effect modification to systems, infrastructure, and knowledge to make the benefits that are important to meet out the smart cities sustainability. The benefits of the smart city are that they make competence. By using well-intended knowledge utensils can advantage administration activities, the atmosphere, and inhabitants to eliminate redundancies,

R. Dhaya
Department of Computer Science, King Khalid University, Abha, Saudi Arabia

R. Kanthavel (✉)
Department of Computer Engineering, King Khalid University, Abha, Saudi Arabia

F. Algarni
Deptartment of Computing and Information Technology, University of Bisha, Bisha, Saudi Arabia

P. Jayarajan
Department of ECE, Sri Krishna College of Technology, Coimbatore, India

A. Mahor
School of EEE, VIT Bhopal University, Bhopal, India
e-mail: amita.mahor@vitbhopal.ac.in

G. R. Kanagachidambaresan et al. (eds.), *Internet of Things in Smart Technologies for Sustainable Urban Development*, EAI/Springer Innovations in Communication and Computing, https://doi.org/10.1007/978-3-030-34328-6_2

find ways to save money, and maintain and manage labor tasks. In the current days applying IoT solutions for smart urban areas and related creative thinking can aid progressive monetary system progression, rising model and status, ascent transportation frameworks, and contour disbursement of visual perception of open resources. The common and vital thing in all smart city-related applications is to proceed with a final conclusion in choosing the decision promptly in obtaining the needed smart and quality accomplishment through machine learning techniques. The best artificial intelligence-based decision-making techniques, machine learning, deep learning, and reinforcement learning algorithms can be used in smart city applications.

2 Preliminary Studies

Relevant research papers in reinforcement learning concepts ministering smart city applications using IoT have been surveyed in various streams that include basics of reinforcement learning and analysis of IoT integration in smart city applications.

Mohammadi et al. [1] proposed a model cum exploration that encompasses deep or profound reinforcement learning to the semi-supervised archetype. As a case learning of smart city solicitations, they absorbed smart buildings and applied the projected exemplary to the problematic of indoor localization centered on BLE sign or signal forte. Joberto S. B. Martins [2] examined the outcome of characterization and man-made consciousness as creative techniques for smart city. An outline encrusted interpretation is also suggested through an assertion near program-based interacting and machine learning effect on oddness chased by a preparation instance that determines the possible guidelines of objective knowledge for smart urban areas.

Zhang et al. [3] proposed a semantic system using IoT with machine learning for smart urban areas. The proposed structure has the ability of acquiring urban information of IoT solicitations on semantic and machine learning progress by taking two contextual investigations, namely contamination finding from vehicles and traffic design presentation.

Streitz et al. [4] presented four viewpoints on the theme stranded in dissimilar methods. First, presenting and imitating on the consequences of the "smart-everything standard," the occasioning proposal trade-offs and their related solicitation to smart cities. Second, deliberating the probable of nonverbal communication for notifying the design of spatial edges for AmI scheme observes is completed. Third, replicating on the part of novel data groups such as "future data" and the role of doubt and their inferences for the next generation of AmI environments is done. Lastly, discussion on the advantages and gaps of the world's major specialized manufacturing communal exertion to expertise a worldwide criterion body on morally associated enterprise for independent and intellectual schemes is done.

Mahdavinejad et al. [5] evaluated the smart urban communities using machine learning techniques by observing information through IoT with the key motto of introducing scientific classification. It also showed how extraordinary strategies are attached to the information so as to take away many data. Relating support vector machine (SVM) on smart city traffic information was also offered for an added thorough examination. Sianaki et al. [6] studied experiments, chances, and investigation tendencies to speak the concerns connected to the statistics era in three businesses of smart cities, healthcare, and transference. All three of these productions might importantly advantage from machine learning and deep learning methods on big data composed by the Internet of things, which is called as the Internet of everything to emphasize the part of linked policies for data collection. The afresh established deep reinforcement learning techniques and their applications in smart cities are also offered and studied.

Kolomvatsos and Anagnostopoulos [7] defined the thought of a query controller (QC) to talk about a smart task for query tasks that engages machine learning by considering two learning plans, namely reinforcement learning (RL) and bunching on investigation. Hammi et al. [8] labeled the current and upcoming inadequacies of smart city and IoT tradition for smart urban communities. Park et al. [9] studied the importance of IoT advancements on the innovation guide investigation for a smart city from their result, the central focuses and vital components for the flourishing betterment of a smart city.

Righetti et al. [10] overviewed the predicted security, social, and moral arrangement in a smart city condition using IoT. Nassa et al. [11] surveyed at shaping the state-of-the-art main technologies, structure, and concern exploited to open ways to direct approaching investigation and to sparkle fresh thought for different industrial projects. Also, they discussed the momentous situation, and possibility lining this tract. Saha and Auddy [12] proposed an idea how Internet and sensors can be utilized to evolve a city to smart by connecting all of these smart characteristics at their forward looking phase of IOT evolution. Patti and Acquaviva [13] depicted the two conveyed IoT stages to change the vitality board in smart city through an approximation of the quantity of vivacity-linked data-presented platforms.

Vijai and Sivakumar [14] deliberated the enterprise of an IoT system based on the study on connected smart responses in the utilization of machine learning systems to smart city which include water request estimation and water quality perception. Mohammadi and Al-Fuqaha [15] proposed a proficient framework to run over the needs of smart city administrations, where the information is classified as labeled information and unlabeled information. The framework uses a combination of labeled and unlabeled statistics to meet the well-regulator policies instead of wasting away the unlabeled data. They similarly explored in what way thoughtful protection education and its adjustment toward semi-supervision can deal with the intellectual side of smart city benefits and modified their execution by outfitting particular utility cases in terms of smart urban communities. They included a few issues just as promising upcoming investigation rubrics for including machine learning and nonstandard state smartness into smart city administrations.

Vijaia and Sivakumar [14] designed an IoT system based on surveys executed on related smart solutions with the psychological side of smart city benefits for different territories of smart urban areas by taking a few issues as prosperous future research rules. Yue Xu [16] concentrated primarily on intelligent machine learning applications meanwhile machine learning harvests IoT and those apparatuses an intelligence to reflect the stretch called embedded intelligence. Neto et al. [17] classified IoT devices as per their abilities, by running machine learning algorithms and covering real doings that formalize the proposed classification. Caminh et al. [18] introduced a smart trust management method, based on machine learning and a flexible transparency window technique that mechanically deals the IoT reserve trust, assessing service provider attributes.

Ruta et al. [19] proposed an agenda for semantic increased data mining on sensor streams, susceptible to resource-constrained inescapable passages. It integrates an ontology-based depiction of statistical arrangements of unusual thinking for fine-grained occasion perception. The characteristic organization difficulty of ML is activated as a source revelation by applying semantic matchmaking. A case study on road and traffic analysis has permitted confirming the suggestion and obtaining an evaluation in connection with the state-of-the-art procedures. Sarker and Sumathy [20] focused on finding the flat of effluence in a particular domain and detect how much more greenery has to be trekked in that zone. With its broad application, they proposed a novel thought with a communal source in order to show the involvement near the atmosphere and manhood to gain agriculturalists and others in contaminated conurbations.

In the above literature survey, it is concluded with the school of thought that the decision-making in all applications of smart city needs further improvement in terms of accuracy for decision-making parts. Thus reinforcement of machine learning could be the best solution for both supervisory and unsupervised part of methods in smart city applications as it incorporated neural network techniques.

3 IoT Technology for Every Smart City

In every smart city applications, sharing information among the components of a network in a faster cum intelligent way to the level of perfection by interconnecting all devices in a seamless manner using IoT.

3.1 Smart Infrastructure

The important properties of smart urban communities, namely smart stopping, smart lighting, and other transportation advancements, are illustrated as follows:

Connected Streets Smart avenues/streets can be fit for getting information and conveying data and administrations to and from endless gadgets, which incorporates data about traffic, street blockages, roadworks for improving open transportation, and the urban perspective.

- **Foot-travelers/Public**: In our cutting-edge world, decompression time is a ware. This space enables individuals to easily occupy a space and to concentrate on collaborating and associating with others.
- **Local Businesses**: Businesses flourish from the acquired intensity of their local economies. By pulling in more individuals into open space and producing valuable, mysterious information, businesses will form new bits of knowledge into their local markets.
- **Municipal/City Activities**: Sensor-empowered information gathering, controllable intelligent lighting, and continuous data enable city governments to more readily comprehend and address the issues of their networks, while guaranteeing their security.
- **Local Schools**: Deliver additional learning chances for the students. Inexpensive connectivity and technology possessions.
- **Property Owners/Landlords**: Upsurge demand for their possessions on the street and thus advanced property values/rents.

Smart Parking and Traffic Management Traffic failures are the foremost issue confronted by residents of a particular place. IoT gadgets can shore massively in giving dynamic and smart arrangements ceaselessly against aggregate traffic troubles and parking spots. The IoT gadgets can profit in sidestepping road-turned parking lots so as to go in rush hour gridlock less occasions and stretch stopping data about stuffed spots. IoT observed traffic signs and can work on vehicle thickness rather than time groups. Stopping sensors can be the best decision in estimating spare time, vitality, and gas outflows and keep up a simple stream of traffic. Identifying free parking spots would definitely reduce traffic congestion in peak hours by means of investigating on electrical car and bike distribution reproductions at urban places. Smart parking management system plays a vital role in determining the empty location for transport at various open spots. Figure 1 shows the top

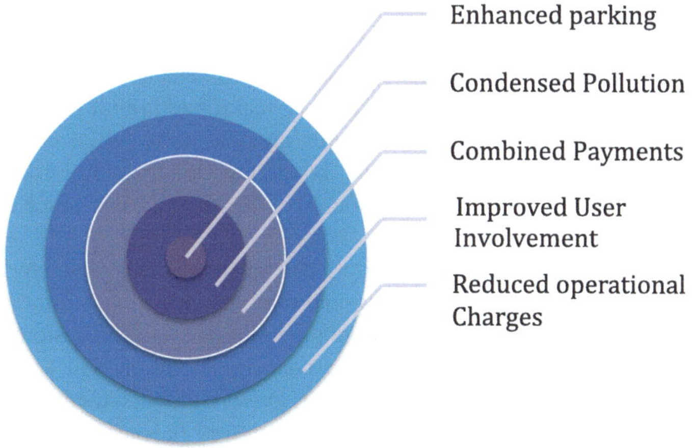

Fig. 1 Top benefits of smart parking

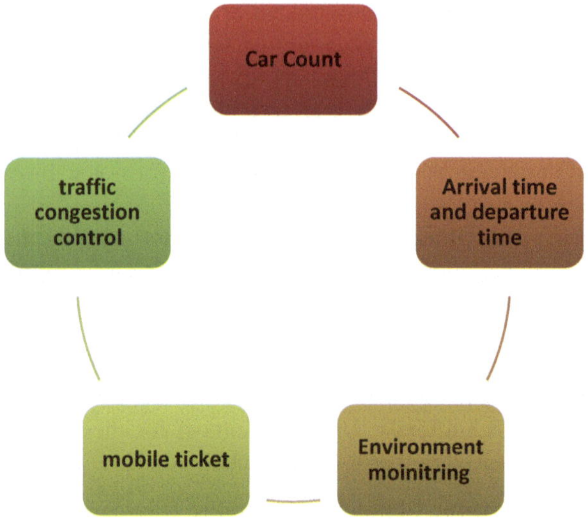

Fig. 2 Traffic management and parking solutions

benefits of smart parking. In ground vehicle spying, sensors are key innovations that give an advanced plan for the drivers to park their vehicles in the crowded places like malls and theaters. The best way is to build remote sensors in parking spots for sending information on the planning and span of the space utilized by means of neighborhood flag processors into focal stopping the board application for smart parking which decreases congestion, vehicle emissions, enforcement costs, and mental strain of the driver. Parking is the groundwork of the smart city. A parking solution can incredibly profit both the client and the owner. Figure 1 shows the major benefits of smart parking. Figure 2 shows the traffic management and parking solutions.

- **Enhanced parking**—With improved parking, users must contact to pursue the constructive parking place which saves time, possessions, and exertion at the same time. It will spring a correct policy to seal a parking lot and space to the profitable and business beings. Henceforth, the road traffic will decrease.
- **Condensed pollution**—Best parking resolution automatically reduces hunt hours, serving in low vehicle productions on a day-to-day basis which eventually decrease the worldwide atmosphere.
- **Combined payments**—Recurring users can substitute everyday, physical cash expenditures through account billing and application of expenditures through their mobile. This could be also allowed the faithfulness of the customer values.
- **Improved user involvement**—A smart parking resolution will assimilate the whole user knowledge into a combined act. Drivers' imbursement, spot identification, location finding, and time announcements all flawlessly develop portion of the endpoint arrival procedure.

- **Reduced operational charges**—Additional automation and fewer physical action protects on work cost and supply tiredness.

3.2 *Waste Management*

In urban areas, the typical issue is to manage the bulk amount of wastes for municipal bodies where people live larger in terms of people. IoT empowered smart bins, trash clearance approaches, observing strategies for trashes would aid hygienic and tidy cum green environment by emerging strategies motorized through renewable energy-based solar-motorized devices to measure the obligatory volume of trash baskets or bins. The process of waste management is discussed in Fig. 3. For waste management systems in the method of biodegradable, e-waste, nondegradable will have a role on saving our environment from further degradation which has minimum effect on our eco-system. A part of smart city is to have a smart water management system and IoT, and associated strategies enable smart water management in the subsequent methods by proper portable aquatic observing to monitor the excellence of tap water in the towns, efficient chemical trickle to identify trickles and wilds of industrial unit in rivers, proper swimming pool remote measurement to control the swimming pool circumstances remotely, effective pollution levels in the sea to control the occurrence of leakages and wastes in the sea, effective water outflows to detect liquid leaking outside tanks and compression differences along pipes, and a smart river flood to monitor water level differences in rivers, dams, and reservoirs.

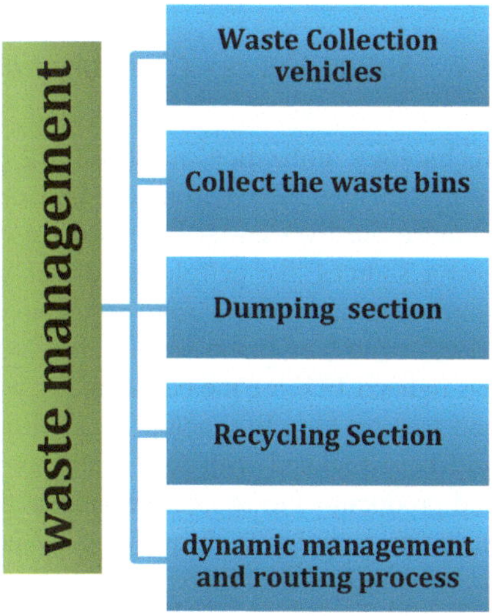

Fig. 3 Process of waste management

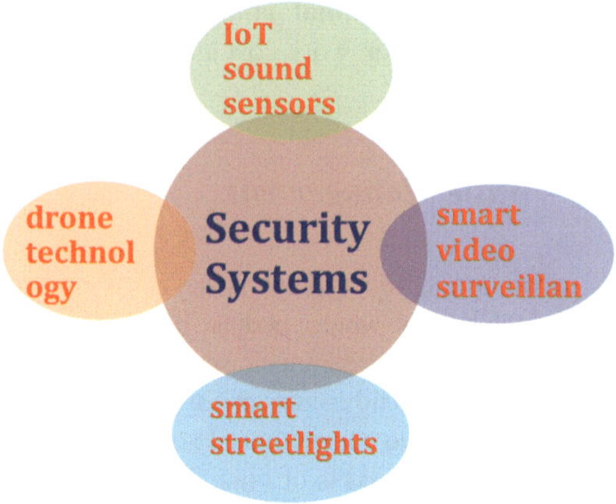

Fig. 4 Security systems for smart city

3.3 Security Systems

The paramount concern of ensuring safety and security in overly crowded towns and cities has been a tough task, and with expertise, it is possible to hint the suspected undesirable rudiments in the forms of detecting the spot of fear and number of shots or strike with individuals by IoT sound sensors, smart video reconnaissance, smart streetlights, and most recent automaton innovation. Figure 4 shows an IoT in security frameworks which can likewise help in proficient observing of open spots like markets, shopping centers, airplane terminals, lodgings, metro stations, banks, and medical clinics. They are absolute necessities in private and business structures.

3.4 Smart Energy Consumption

Technology utilization must on the one hand be at its peak level without compromising the environment and on the other hand not affect the ecosystem to make our lives easier by making use of characteristic assets as non-sustainable power sources like petroleum, diesel, coal, and wood. To protect the fate of our coming ages, nations classified resources into sustainable type of vitality like sun-based, wind-based, and water-based. Instead of using lights like tungsten based or gas lights, LEDs are the best choices, which do have the capability of reducing cost and best on long life factor. In addition, the novel procedure of energy ingestion will be associated with IoT strategies which initiate serving folks, public forms, and productions to square the vitality metrics. In addition, the act of saving water ingesting cumulative air quality, sewage disposal, and active power generation are the essential ones. Figure 5 shows the types of energy sources like wind energy, renewable energy, thermal energy, nuclear energy, and hydraulic energy.

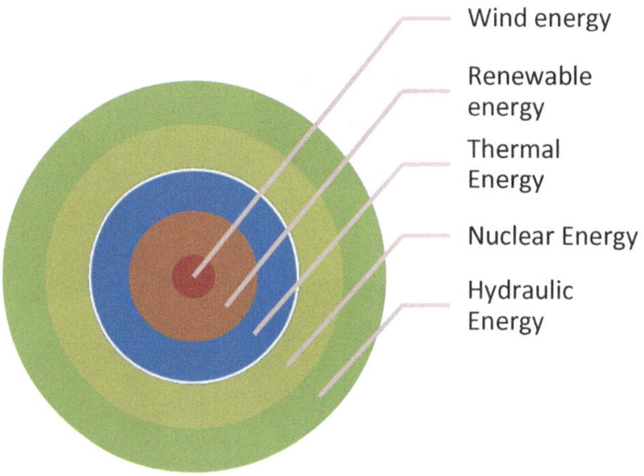

Fig. 5 Types of energy sources for smart city

3.5 Healthcare Services

The desk-bound lifestyles of modern age have resulted different classes of illnesses and fitness difficulties in urban populace irrespective of a progression in medication and knowledge that enabled lifetime anticipation of humanoid existences. IoT in well-being segment assistances in isolated observing, clever sensors, and movement follower gadgets where smart urban communities want smart medical clinics, which can follow patients remotely. To the other side from restorative application, the production of viable development through better use of sustenance, vitality, medical clinic squander board, stock administration, and so on, future smart urban communities will be with legitimate well-being and health centers for their inhabitants.

4 Smart City Paradigm and Its Applications Using IoT

The smart city paradigm illustrated in Fig. 6 shows how the essential smart city has been interconnected with IoT applications effectively to sustain the surroundings cleanly and greenly without affecting the ecosystem to make the human life comfortably.

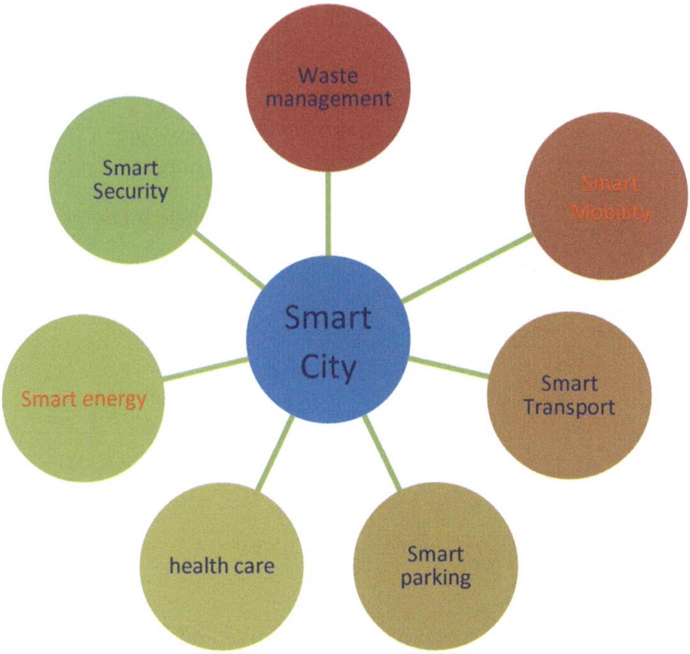

Fig. 6 Smart city paradigm and IoT applications

4.1 Smart Energy

Since smart energy is one of the utmost significant investigation parts of IoT, it is imperative to reduce the general influence depletion to provide superior and reasonable ecologically approachable energy. The smart grid is one of the greatest vital requests of smart energy which comprises numerous high-speed time sequence statistics to observe important plans. Furthermore, the smart energy substructure would develop extra multifaceted approaches in the future to optimize energy resource consumption.

4.2 Smart Mobility

The important aspect of any city is keeping smart mobility that plays an important role to progress the excellence of lifetime in the conurbation. Smart movement can be alienated into the subsequent three main mechanisms:

- ***Autonomous cars***: IoT plays a main role in identifying the belongings in what way vehicles are ridden in order to improve vehicle services with the help of IoT sensor devices and wireless networks which makes it probable to generate self-driving vehicles and observe vehicle concert. By smart applications of IOT, the traffic spots can be detected to avoid running on that optimized run way that enables to improve security and safety of the users by collecting data from cars, road cameras, and security sensors connected on highways or roads.

- ***Community transportation***: IoT can also be used to progress the community transportation scheme managing through a precise location and steering information to a smart transportation scheme.

4.3 Smart Citizens

This training circumstance for smart cities shelters an *extensive* variety of zones of humanoids, such as conservational observing, corruption observing, and communal fitness. The atmosphere with all its apparatuses is essential and vigorous for lifetime. Subsequently, constructing development in knowledge is assured to enrich safety. Neighboring watching keenly to criminality would else subsidize to complete communal fitness.

4.4 Urban Planning

As the conurbation and atmosphere together show foremost parts in humanoid lifespan, getting verdicts in this background is vital and time-consuming one. By accumulating statistics from dissimilar causes, it is conceivable to mark a verdict for the future of the metropolis. Taking resolutions distressing the metropolis substructure, plan, and functionality is termed urban planning. IoT is helpful in this zone; because of the finished smart city statistical investigation, the experts could envisage which portion of the conurbation will be extra congested in the future and catch the explanations for the possible difficulties. Amalgamation of IoT and urban planning will require a foremost consequence on preparing imminent substructure expansions.

4.5 Smart City Data Characteristics

Smart city maneuvers produce statistics in an uninterrupted way, demonstrating that the statistics congregated from traffic, health, and energy management applications would engender a substantial volume. Tallying, since the data generation rate fluctuates for dissimilar maneuvers, meting out statistics through diverse group charges is an experiment. Table 1 explains the smart city paradigm and IoT applications.

Table 1 The smart city paradigm and IoT applications

Applications	Uses
Smart energy	It is imperative to reduce complete influence feasting to provide excellent and reasonable environmentally approachable vitality
Smart mobility	Keeping smart mobility that plays an important role to progress the excellence of lifespan in the metropolis
Smart citizens	Shelters an extensive variety of zones in hominid lives, such as ecological observing, corruption observing, and communal fitness
Urban planning	Taking decisions distressing the conurbation infrastructure, plan, and functionality
Smart city data characteristics	Produce statistics in an uninterrupted way, demonstrating that the statistics congregated starting traffic, health, and energy management requests would engender a substantial volume

5 Machine Learning Concepts in Smart City

Data is the essence of the modern city being captured by sensors generating massive streams of data that needs analysis. Artificial intelligence (AI) is the main technique to transform this information into understanding by means of capture, inspect, and analyze data to influence the whole thing from traffic and parking administration to law implementation and city amenities.

Massive Data Management The majority quantity of data formed by IoT devices can be enormous and rough to achieve in positions of development, communication, or transmission and storage by following lack of standardization and interoperability as the smart city must hold three core values. Figure 7 is shown a conceptual construction of a deep reinforcement learning scheme.

- **Livability:** Cities that must offer hygienic, strong living environments deprived of contamination and blocking through a cardinal substructure that brands conurbation facilities promptly and suitably obtainable, whatever and wherever.
- **Workability:** Metropolises must deliver the allowing substructure in the form of energy, connectivity, and computing, essential service to strive internationally for great value jobs.
- **Sustainability:** Metropolises that arrange for facilities deprived of pocketing from upcoming compeers and of course learnability, the impartial admission to quality learning chances.

Big data and inexpensive computing also empowered the bang of artificial intelligence (AI), program that absorbs in software applications subsidiary every single characteristic of lifespan. Enhance aiding utensils similar to robotics, drones, and independent automobiles as exposed in Fig. 8 to make the smart city out of the complexities.

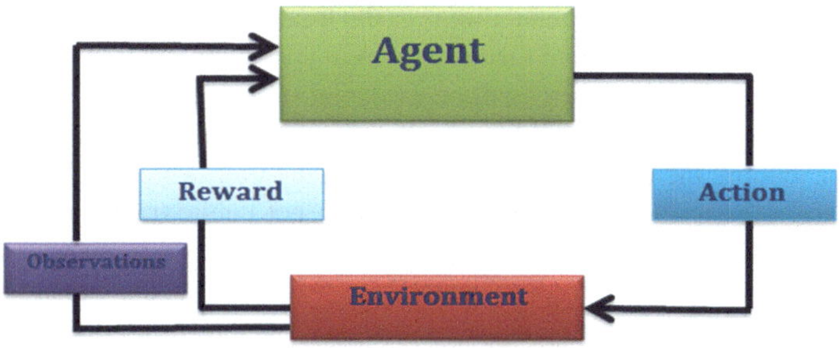

Fig. 7 An intangible construction of a deep reinforcement learning system

Fig. 8 Tools to transformation of smart city

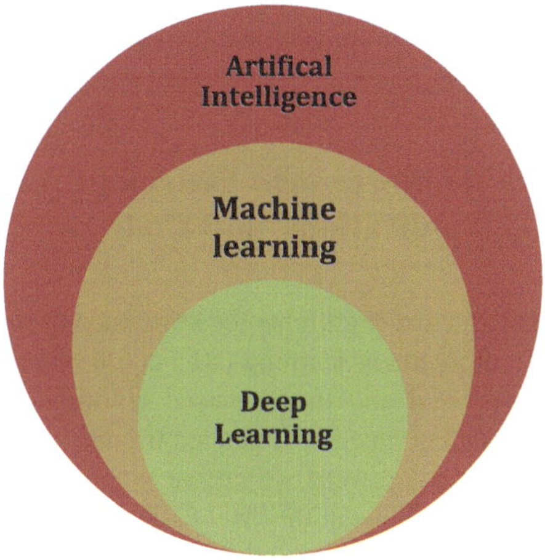

AI solicitations are currently predominant in healthcare for identifying syndrome and enlightening public healthiness, transport (traffic control), community security by facial appreciation, manufacturing in terms of process control, and connected promotion. The nine immense hard-hitting challenges directing in the smart cities are unemployment, income inequality, privacy, algorithmic bias, access, machine ethics, automation, humanity, and genome editing.

6 Big Data and IoT Changing a City into a Smart City

There are various conurbations in the world which are measured smart because they have accomplished convinced parameters such as traffic and environment management.

- The garbage containers should have a distinct card-based interface, and they will grind only when chip cards are recycled to contact them. The generated data help the planners to successfully define the disposal pattern and advocate solutions and also will carry the unwanted things to the treatment plants by sucking it from the containers as garbage and then use them in energy production. Thus, the part of air-polluting garbage trucks will be excluded.

- The RFID tags in all cars will help in the effective management of traffic in the form of sending the data to the central monitoring unit, which will investigate it to progress the traffic system. People can also get public transport information instantly on their smartphones at all times.
- Smart street lamps measured by a central power grid can regulate their brightness conferring to the sum of people in the area which will allow the city to protect energy by switching off some lamps in the spaces where low population attentiveness is noticed.
- Data mining will boost security measures, and kids will have distinct bracelets to find their accurate site during tragedies.

7 Reinforcement Learning: An Introduction to the Concepts, Applications, and Code

This section highlights the key concepts and terminology in reinforcement learning. Reinforcement learning (RL) could be observed as a method which derives among supervised and unsupervised learning. It is not harshly supervised as it is not contingent on a set of labeled training data but is not unsupervised learning because we have a reward which we want our agent to make the most of it. The agent requests to invent the "right" deeds to yield in diverse circumstances to accomplish its whole objectives. Reinforcement learning is the technique of better decision-making. Reinforcement learning encompasses no supervisor, and only a reward signal is used for an agent to regulate if they are performing well or not. *Time* is a key component in RL where the process is consecutive with delayed feedback. Each action the agent marks disturbs the next data it obtains.

7.1 Reinforcement Learning Problem

The important task in reinforcement learning technique is that the agent needs to detect the decision by means of reward signal. The reward R is a scalar feedback signal which designates how fit the agent is performing at step time t. In reinforcement learning, it is a need to define the problem such that it can be implemented to satisfy the phenomena called "reward hypotheses." Reward is like a prize for winning in a game and punishment is nothing but to lose a game. Here all goals can be labeled by the maximization of estimated cumulative reward. Since the process includes sequential decision-making tasks, the actions to sort out primarily

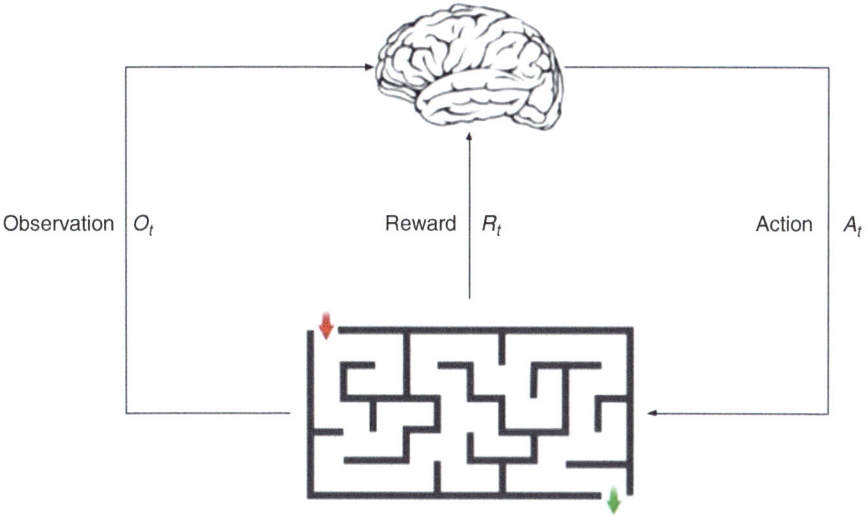

Fig. 9 Reinforcement learning process

on may have a long-term consequence on the overall *goal*. Sometimes it may be better to compromise immediate reward to gain more long-term reward.

7.2 Background Awake Reinforcement Learning Problems

Reinforcement learning process diagram is shown in Fig. 9. In reinforcement learning, the **agent** receipts the decisions on which *schedules* to yield at each time step *A*. The agent makes these decisions centered on the scalar *reward R* it receives and the observed environment *O*.

The environment obtains the exploit from the agent and harvests a new observation *O* and scalar reward *R*. The things happen next to the environment depends on the history. The previous *H* is a categorization of explanations, actions, or activities and payments up to time *t*.

$$H_t = A_1, O_1, R_1, \ldots A_t, O_t, R_t \tag{1}$$

where

H_t sequence of observations
O_t new observations
R_t reward

The state is the information for defining the belongings to ensure the next.

$$S_t = f(H_t) \tag{2}$$

where S is the state.

The important difference amid the *history* and *state* is that a function of the history is known as state. The states can be divided into three key types:

- ***Environment State*** (S^e)*:* The environment's private mark and might not be noticeable to the agent which is rummage sale to pick the next observation.
- ***Agent State*** (S^a)*:* The agent's interior illustration and is rummage sale by the agent to choice the next exploit.
- ***Information State/Markov State*** (S)***:*** Includes helpful information of the history and henceforth assumes that for this state it will be suitable evidence to perfect the future, and the history can be shown out.

Markov State A state S is Markov if and only if

$$P\left[S_{t+1}|S_t\right] = P\left[S_{t+1}|S_1, \ldots S_t\right] \tag{3}$$

The condition is likewise isolated into completely observational condition and partially observational condition.

- ***Fully Observable Environments*** *(Markov Decision Process):* Agent straightly detects environment state.

$$O = S^{\mathrm{a}} = S^{\mathrm{e}} \tag{4}$$

- ***Partially Observable Environments*** *(Partially Observable Markov Decision Process):* Agent indirectly detects environment.

$$S^{\mathrm{a}} \neq S^{\mathrm{e}} \tag{5}$$

7.3 Reinforcement Learning Agent

The description to set up RL and in what way the decision-making procedure has been well-defined. An RL agent can have one or other of the three major mechanisms specifically policy, value function, and model as stated below.

- ***Policy:*** Agent's behavior function which is a plan from state to action, and it can be a *deterministic policy* or a *stochastic policy.*

  Deterministic policy function:

$$a = \pi(S) \tag{6}$$

Stochastic policy function:

$$a = \pi(s) = P\left[A_t = a|S_t = s\right] \tag{7}$$

- **Value Function:** Epitomizes exactly how worthy every state and/or action is a prediction of future compensation.

$$v_\pi(s) = E_\pi\left[R_{t+1} + \Upsilon R_{t+2} + \Upsilon_2 R_{t+3} + \cdots |S_t = s\right] \tag{8}$$

Sample value function

- ***Model:*** Agent's depiction of the situation, which predicts the situation to be followed. The predictions are of the following state and subsequent instantaneous reward.

Sample equivalence to forecast of the following state.

$$\boldsymbol{P}^{\mathbf{a}}_{\mathbf{ss'}} = \boldsymbol{P}\left[S_{t+1} = s'|S_t = s, A_t = a\right] \tag{9}$$

Sample equivalence to forecast the next instantaneous reward.

$$R^{\mathrm{a}}_{\mathrm{s}} = \boldsymbol{E}\left[R_{t+1}|S_t =, A_t = a\right] \tag{10}$$

RL agents can be well-defined into the following kinds:

- **Value Based:** Consumes no policy, picks actions voraciously founded on state values.
- **Policy Based:** Consumes no value function, uses the policy function to pick actions.
- **Actor Critic:** Practices together value and policy functions.
- **Model Free:** Practices policy and/or value functions but takes *no* model.
- **Model Based:** Practices policy and/or value functions and takes a model.

7.4 Problems within Reinforcement Learning

The two fundamental problems in sequential decision-making are as follows:

- ***Learning:*** The environment is originally *unknown*, and the agent needs to interrelate through the environment to progress its policy.
- ***Planning:*** If the environment is *recognized*, the agent achieves calculations through its model and then progresses its policy.

Since reinforcement learning is similar to an experimentation learning technique, the specialist needs to gain from its understandings in the earth to make choices without dropping an excess of remuneration. It needs to think about misuse and investigation with a harmony between investigating new data and utilizing realized data to amplify compensate.

8 Optimizing Traffic Using Reinforcement Learning

As urban populaces goes on for a long span to create urban areas from spot to put on the planet will fundamentally change the setup to experience their normal business and eco-accommodating necessities. Less than impressive transport systems can influence the earth through air contamination, clamor, and residue. Modifying these systems can tip to a reduction in petroleum derivative admission as voyage times are abbreviated and more advantageous air quality for natives. Thus, solid transport arranges progression demonstrates a noteworthy job in the urbanization and maintainability of the urban communities. Fundamentally, the beginning of the urban traffic conveyance issue is in multicriteria-based leadership. The reinforcement learning structure, in which a specialist obtains from a run of the mill with ideal approach dependent on its condition, could bear the cost of an advantageous procedure for algorithmic development and system upgrade. Every accomplishment that the specialist would income will prompt a recompense or discipline with the new perception of the state. Over its learning development, the specialist will contemplate a conveyed directing approach that might take advantage of the limit of the urban transport arrangement. This procedure could be saved as a Markov decision process (MDP), which inevitably provides the greatest answer by increasing point-by-point arrangement well ordered.

A Markov decision process comprises five key fundamentals as revealed underneath:

- *S*: The group of the states will possibly reproduce the situation with detailed possessions.
- *A*: A set of actions might modify the position.
- *P*(*SA*): The possibility of the state *S* varies below action *A*.
- γ: The discount feature.
- *R*(*SA*): The expected recompense that state *S* modifies below action *A*.

The perspective of the urban transport network is a fairly expedient and composite framework (the choice tree is difficult to assess, and numerous choices must be set aside a few minutes), and it would be utilitarian without model reinforcement learning strategy. For this situation, by practice Q-learning is done through capacity estimation to get familiar with the best vehicle circulation technique for viable and solid urban transportation.

Q-learning calculation is a perfect allowed reinforcement learning arrangement ready to get an activity determination strategy for MDP. *Q*-learning calculations crush with a general MDP as named below. The Bellman condition (Eq. 1) given below extends the ideal strategy of secure progress likelihoods and accolades:

$$\mathbf{Q}\left(\mathbf{s},\mathbf{a}\right)=\mathbf{R}\left(\mathbf{s}'\right)+\gamma_{\mathbf{max}}\mathbf{Q}\left(\mathbf{s}',\mathbf{a}'\right) \tag{11}$$

$Q(s,a)$ is the rapid compensation $R(s')$ for manufacturing the movement in accumulating to the finest usefulness Q for the ensuing state $Q(s', a')$. For all state and activity pair Q(s, an), esteem is instated to zero. At that point watch the current states' by recounting the accompanying advances:

- Elite an action and understand it.
- Accept instantaneous compensation R.
- Eyewitness fresh state' s'.
- Inform $Q(s, a)$ by the calculation overhead with the discounted reward γ.
- Set $s = s'$.

The Bellman condition assumes an essential job in reinforcement learning as it is broadly connected amid approach appraisal. On the off chance that the Q capacity can be effectively evaluated, a superior approach $\pi(s)$ will build up the ideal strategy and lead the activities as per Eq. (12).

$$\pi\left(s\right)=\mathbf{arg\ max}\ \boldsymbol{Q}\left(s,\boldsymbol{a}\right) \tag{12}$$

Through the Q-learning method, a combination of key potentials could be separated, with the inducement for the state highlights, remunerate. The following state highlights once it cooperates in recreation.

The essential capacity is to decrease the all-out foldaway period of each driver in an established urban conveyance. In this manner, it could prompt consolidated carbon dioxide outflows and declined non-renewable energy source utilization. The remuneration of this procedure is developed since the allowed volume moderates the complete foldaway period of the congregation of vehicles in the urban transportation assemble. The activity, however, is fairly unobtrusive; it just includes way determination. With a foreordained dimension of normal hold up times, the operator will pick which vehicle to allot an entitled way for gathering course improvement. The state highlights the properties that show the state of the traffic. It needs to mirror the bustling dimension of the urban transportation arrange. So as to accomplish this, a neural system may be connected to "read" the circumstance of the transportation organizes. By blending reinforcement learning and a neural system, the capacity to oversee the circumstance of the vehicle organize and succeed the traffic could be improved as needs be. Reinforcement learning is a favorable illumination to this trouble as it allows various prospects to circulate the ideal course. With certain activities, the condition of the vehicle organization can be modified, from which we can decide whether to remunerate or rebuff the calculation with right properties. This is commendable for dynamic control and is essentially proposed for the alteration of various situations. Keen administrations are a fundamental area of understanding urban areas and the biological bases of Internet of Things (IoT) wherever the vision overdue the societies is clutched and enhanced over the tactile information. Up to a cumbersome amount of preparing information isn't consistently sensible; thus, there is a need to consider surprising techniques that incorporate unlabeled information too. Reinforcement learning (DRL) has included incalculable achievement in various accommodation spaces. As a contextual investigation of keen city applications, centers around savvy structures must be connected, and indoor confinement is the preeminent module of shrewd city administrations since individuals invest critical energy in indoor situations.

9 Conclusion

This chapter has presented the reinforcement learning concepts ministering smart city applications using IoT. The initial section of the chapter focuses on the smart city applications and its interconnection with the IoT. The later section explained about the deep learning and reinforcement learning concepts related to the smart city developments. In the majority of the cases, the basic issue for open experts (typically, neighborhood, similar to regions) is the proficient administration of information towards the help of novel administrations. The reason is that examinations given over the gathered information will possibly care in the carriage of fresh submissions that resolve encourage natives' lives. Be that as it may, the arrangement of examination requests insightful systems for the hidden information the board. The most realized strategy is the detachment of enormous volumes of information into various portions and its equivalent management to constrain the essential period aimed at the conveyance of investigation. In this chapter, the concept of reinforcement learning has explained the reward and punishment with regard to the agent. It has also explained an intellectual procedure intended for inquiry obligations that receives machine learning (ML). The goal of bringing a well-organized frame to overhaul the decision production of the information in any game or task with more complexities using mathematical function has been done logically. Over a comprehensive investigatory assessment, this chapter has discloses the benefits of the proposed models using reinforcement learning techniques that played a vital role in efficient decision-making.

References

1. Mohammadi, M., Al-Fuqaha, A., Guizani, M., & Oh, J.-S. (2018). Semi-supervised deep reinforcement learning in support of IoT and smart city services. *IEEE Internet of Things Journal, 5*(2), 624.
2. Martins, J. S. B. (2018). Towards smart city innovation under the perspective of software-defined networking, artificial intelligence and big data. *RTIC - Revista De Tecnologia Da Informação E Comunicação, 8*(2).
3. Zhang, N., Chen, H., Chen, X., & Chen, J. (2016). Semantic framework of internet of things for smart cities: Case studies. *Sensors, 16*(9), 1501.

4. Streitz, N., Charitos, D., Kaptein, M., & Böhlen, M. (2019). Grand challenges for ambient intelligence and implications for design contexts and smart societies. *Journal of Ambient Intelligence and Smart Environments, 11*, 87–107.
5. Mahdavinejad, M. S., Rezvan, M., Barekatain, M., Adibi, P., Barnaghi, P., & Sheth, A. P. (2018). Machine learning for internet of things data analysis: A survey. *Journal of Digital Communications and Networks, 20*, 1–56.
6. Sianaki, O. A., Yousefi, A., Tabesh, A. R., & Mahdavi, M. (2019). Machine learning applications: The past and current research trend in diverse industries. *Inventions, 4*, 8. https://doi.org/10.3390/inventions4010008.
7. Kolomvatsos, K., & Anagnostopoulos, C. (2017). Reinforcement learning for predictive analytics in smart cities. *Informatics, 3*, 16. https://doi.org/10.3390/informatics4030016.
8. Hammi, B., Khatoun, R., Zeadally, S., & Fayad, A. Internet of things (IoT) technologies for smart cities. *IET Networks, 7*(1), 1–13. https://doi.org/10.1049/iet-net.2017.0163.
9. Park, E., del Pobil, A. P., & Kwon, S. J. (2018). The role of internet of things (IoT) in smart cities: Technology roadmap-oriented approaches. *Sustainability, 10*, 1388. https://doi.org/10.3390/su10051388.
10. Righetti, F., Vallati, C., & Anastasi, G. (2018) IoT applications in smart cities: A perspective into social and ethical issues. In *2018 IEEE International Conference on Smart Computing (SMARTCOMP)*. https://doi.org/10.1109/SMARTCOMP.2018.
11. Nassa, A., Montasser, A. H., & Abdelbaki, N. (2018) A survey on smart cities' IoT. In *International Conference on Advanced Intelligent Systems and Informatics*. https://doi.org/10.1007/978-3-319-64861-3_80
12. Saha, H. N., & Auddy, S. (2017). IoT solutions for smart cities. In *Industrial Automation and Electromechanical Engineering Conference (IEMECON), 2017 8th Annual, At Bangkok, Thailand*.https://doi.org/10.1109/IEMECON.2017.8079565.
13. Patti, E., & Acquaviva, A. (2016). IoT platform for smart cities: Requirements and implementation case studies. In *2nd IEEE International Forum on Research and Technologies for Society and Industry Leveraging a better tomorrow (RTSI 2016), At Bologna, Italy, Volume: Proceedings, 2016*. https://doi.org/10.1109/RTSI.2016.7740618
14. Vijai, P., & Sivakumar, P. B. (2016). Design of IoT systems and analytics in the context of smart city initiatives in India. *Procedia Computer Science, 92*, 583–588.
15. Mohammadi, M., & Al-Fuqaha, A. (2018). Enabling cognitive smart cities using big data and machine learning: Approaches and challenges. *IEEE Communications Magazine, 56*(2), 94–101.
16. Xu, Y. (2015). *Recent machine learning applications to internet of things (IoT)*. Retrieved from http://www.cse.wustl.edu/~jain/cse570-15/index.html
17. Neto, A. R., Soares, B., Barbalho, F., Santos, L., Batista, T., Delicato, F. C., et al. (n.d.) Classifying smart IoT devices for running machine learning algorithms.
18. Caminha, J., Perkusich, A., & Perkusich, M. (2018). A smart trust management method to detect on-off attacks in the internet of things. *Security and Communication Networks, 2018*, 6063456.
19. Ruta, M., Scioscia, F., Loseto, G., Pinto, A., & Di Sciascio, E. (2019). Machine learning in the internet of things: A semantic-enhanced approach. *Semantic Web, 10*(1), 183–204.
20. Sarker, D., & Sumathy, S. (2017). Cognitive IoT incorporating intelligence in building smart environment. *IOP Conference Series: Materials Science and Engineering, 263*, 042012. https://doi.org/10.1088/1757-899X/263/4/042012.

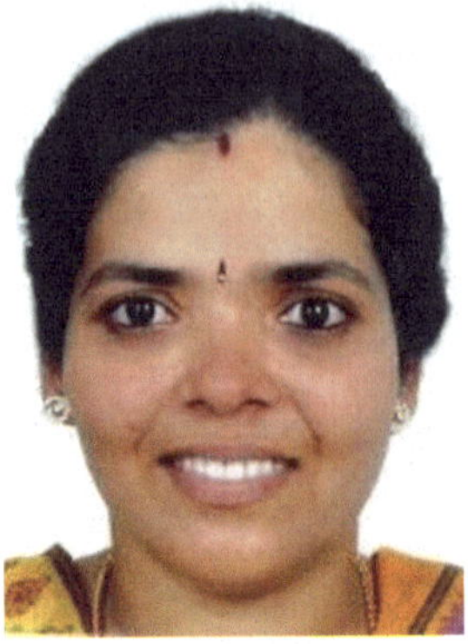

R. Dhaya is presently working as a faculty in the Department of Computer Science, King Khalid University, Kingdom of Saudi Arabia. She has published books on microprocessor and microcontroller, fundamental of computer programming, multi-core architecture, software architecture, mobile computing, information retrieval and real-time embedded systems as a proof of her academic excellence and expertise. She is a lifetime member of professional societies of IEI, IETE, ISTE, IAENG, and IACSIT. She also received Young Engineer Award in 2017 from the Institute of Engineers, Kolkata, India. She has published more than 50 research articles in peer-reviewed journals and 85 international conferences. Her research interests are sensor networks embedded systems, DSTN, cooperative communication, and cloud computing.

R. Kanthavel has been in the teaching field for the last 20 years holding various positions as lecturer, assistant professor, professor, dean, vice principal, and principal in government and other reputed institutions. He has published more than 60 research articles in peer-reviewed journals and 110 international conferences. His research interests are sensor networks, embedded systems, cooperative communication, cloud computing, big data, IoT, and image processing. He has published technical books on microprocessor and microcontroller, fundamentals of computer programming, multi-core architecture, software architecture, mobile computing, electromagnetic theory, and social network analysis. He has also acted as resource person in the form of session chair, chief guest, and keynote speaker in many national- and international-level workshops, seminars, and FDPs. He is a lifetime member of professional societies of IEEE, CSI, IETE, ISTE, and IACSIT.

Fahad Algarni is the dean of the College of Computing and Information Technology at the University of Bisha, Saudi Arabia. He received his PhD from Clayton School of Information Technology, Monash University, Melbourne, Australia. He received his bachelor's degree from the Department of Computer Science at King Abdul-Aziz University with First Class Honors. He also graduated with master's degree in computer networks from La Trobe University, Melbourne, Australia. He has been in the teaching field as academician, administrator, and researcher for the last one decade. He has published many research articles in peer-reviewed journals and international conferences. His research interests are Internet technologies, networks, green computing, IoT, E-services, and big data analysis.

P. Jayarajan has completed his BE (EEE) from Madurai Kamaraj University in 2004, ME (Applied Electronics) from Anna University in 2008, and PhD in the field of wireless sensor network from Anna University. He has about 11 years of teaching experience and presently working as an associate professor in the Electronics and Communication Engineering Department, Sri Krishna College of Technology, Coimbatore. He has published 15 papers at international journals and conferences. His research interest includes wireless sensor network, modeling and simulation, and IoT.

Amita Mahor has completed her BE (EEE) from Jiwaji University, Gwalior in 1997, ME (Heavy Electrical Equipment) from Maulana Azad National Institute of Technology, Bhopal in 2006, and PhD in the field of optimum generation scheduling in 2010. She has about 20 years of teaching experience at various levels and presently working as Professor and Dean of School of EEE, VIT Bhopal University, Bhopal. She has published 70 papers at international journals and international conferences. Her research interest includes application of evolutionary computational techniques in electrical engineering, power electronics, and power system.

Internet of Things Based Solutions and Applications for Urban Planning and Smart City Transportation

A. Suresh, R. Udendhran, and M. Balamurugan

1 Introduction

IoT is said to be the computationally augmented environment around us that supports the integration of physical world with the information world, which depends on the ever-growing infrastructure of the internet through the active range over large area and certain other smart sensors that include GPS, RFID, infrared sensors, laser scanners, NFC, and accelerometers. IoT is defined as "the interconnection among the exclusively detectable embedded devices, virtual, physical, and smart objects [Interconnections made by targeting the environment as well as serves the best advantage to the humans] with the help of the Internet Protocol version 6 (IPv6) (internet infrastructure that has unlimited capability), communication modules, embedded systems, and sensing-actuation capabilities" [1].

IoT serves in almost every application and promotes a fascinating construction with the more sophisticated network, and the objects that are physically connected could include railways, electric network, machines, goods, roads, animals, appliances, plants, water system, construction, etc. It could be finely said that the connectivity could be made among the machines, people, and places [2]. IoT technology integrated along with the Artificial Intelligence (AI) seems to unlock huge opportunities in various fields that include adaptive energy systems, autonomous vehicle, and smart healthcare [3]. These examples are considered to be an existing one and still the opportunities with the integration of IoT and AI grows tremendously in all possibilities including the industries and the consumers [4]. But

A. Suresh (✉)
Department of Computer Science and Engineering, Nehru Institute of Engineering and Technology, Coimbatore, Tamilnadu, India

R. Udendhran · M. Balamurugan
Department of Computer Science and Engineering, Bharathidasan University, Trichy, India

G. R. Kanagachidambaresan et al. (eds.), *Internet of Things in Smart Technologies for Sustainable Urban Development*, EAI/Springer Innovations in Communication and Computing, https://doi.org/10.1007/978-3-030-34328-6_3

this combined technology is not found everywhere around the cities although the efficiency of AI along with the IoT nexus is readily available with the deployment of hardware and software. Considering the fact of sustainability, scalability, and impact factors of integrating AI along with IoT, significant collaboration and planning is required. In the case of multiple ownerships, all partners are taken into account for collaboration and planning [5].

The control could be lost by the cities to the academic institutions, private companies, and even citizens in the case where the overarching digital master plan could be confirmed by any city authorities since the solution offered by the IoT is emergent. The question always arises in a way whether the sensors owned by the academia, cities, or the business have the computational power in collecting the raw data, filter out the necessary data by applying certain algorithm based on the machine learning process? If this process is done, then what will be the necessity for the process to be carried out?

More commonly, the sensor installed by these entities involves in collecting the citizens and the infrastructure data either by the approval explicitly. The performance based on the outline of the citizens could be done through AI depending on the data they offer that could promote advancements in the cities and also the citizens or either it could be utilized for the profit of the corporate. Certain barriers should be crossed in order to succeed in achieving the deployments of the cities. Commonly, the only target is to inform AI through the data gathered by the IoT in the real-time process. At certain times, the generation of data is the most main part that is done by several elements and this could be employed in a deliberate way. Certain challenges are also faced in the urban service sectors (transportation, healthcare sector) by promoting AI with the IoT.

For example, consider city x that consists of ten lampposts traffic counting sensors. In this case the physical sensors are being manufactured by company A and the other company B provides the edge computing software and company C promotes the hosting and centralized data management. The planning permission is granted by the Local Authority Y and certain politicians Z based on the analytical decisions.

The strategy should be defined with the use-cases so that the measurements could be customized, planned and considered before the deployment process takes place. Without this strategy, the output will be pool of raw data with the other business content whose use will be a great challenge [6]. The areas that deal with the security, privacy, ethics, and legality form as a core field with the higher priority rather than the technical insight. Moreover, the city authority could get benefit from the machine learning along with the AI. For instance, reduction of accidents due to the traffic and enhancing the transportation and quality of air are the high demands by the agenda. These challenges could be answered when the right information with the intelligence based goes into the hand of the empowered authorities and the necessary actions could be taken by them. IoT generates significant and necessary data and with the AI algorithm the data could be considered at the sufficient volumes and makes use of the AI algorithms [7, 8]. For the authorities to act on the AI/IOT nexus, proper impact should be necessarily considered, and the power should fall on to the

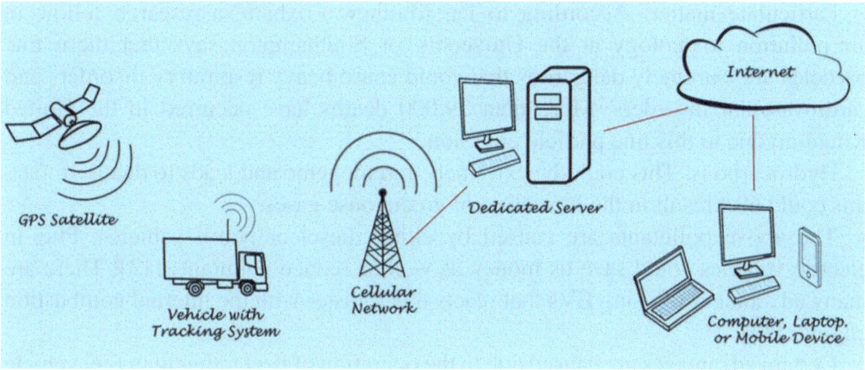

Fig. 1 IoT in smart fleet management

right hands. In the context of cities, change of policy will necessarily derive the values or potential decision could be made by the independent decision maker. For instance, IoT is integrated in fleet management for enhanced performance as shown in Fig. 1.

Citizen-focused and test-driven approaches make sure that the end scenario is one that has been tested with the public along the way, and not just imposed upon them. After this significant experimentation, the theories could be tested with the stakeholders and then implementing it in the small scale without any huge investments [9]. This could give a massive lead along the technology as well as the techniques adapted and further deals with the effective discussion of citizens [10].

2 Gearing Up for IoT-Based Electric Vehicles for Urban Planning and Smart City

The survey taken at the year 2016 states that 17% of the diesel vehicle has lost its sales due to the extreme pollution that could be hazardous when compared to the petrol vehicle [11]. The toxic gases that are emitted by these vehicles are as follows:

Carbon monoxide (CO): Oxygen plays a key role in our body. With the inhalation of carbon monoxide it could block the oxygen carrying capacity that could affect the most important organs of our body. Extreme emission of CO could be dangerous and should be controlled. But a limited emission could save the normal people but could affect those people suffering from the heart disease.

NOx (oxides of nitrogen including nitrous oxide and nitrogen dioxide): This could extremely affect the lung functionality causing illness in breathing. Extreme exposure to these gases could cause asthma and other allergies to the lungs. Nitrogen dioxide (NO2) could affect the environment that could cause acid rain and ozone depletion [12].

Particulate matter: According to Dr. Matthew Loxham, a research fellow in air pollution toxicology at the University of Southampton says that these fine particles are extremely dangerous that could cause heavy respiratory disorders and cardiovascular disorders. More than 29,000 deaths have occurred in the United Kingdom due to this fine particle emission.

Hydrocarbons: This could be extremely carcinogenic and leads to death. In fact, this could also result in the formation of greenhouse gases.

The above pollutants are caused by either diesel or petrol vehicles. Plug-in electric vehicles could save us money as well as reduce pollutants [13]. There are many advantages of using EVs that reacts much faster with the internal combustion engines [14].

Certain advantages are caused due to the operation of fleet rather than fuel vehicle [15]:

- Electric vehicles (EVs) react quickly. It has a very good torque and extremely responsive.
- EVs charging is user-friendly. Just like plugging a smart phone the vehicles are connected. Moreover, we can able to monitor the amount of charge taken by our vehicle.
- The cost of the fuel is highly controlled using IoT-based electric vehicles since we could charge the vehicle using any form of renewable sources from solar and wind.
- Electric train for smart urban transport: Electric trains are environmental friendly when compared [16]. Since the components of the electric vehicle are reduced say for example, it has only electric motor. They do not require any replacement of fuel. In fact, features like regenerative braking reduce wear and tear on basic vehicle components. Recent survey says that a typical fleet that uses electricity reduces 50 percent of its cost for the maintenance purposes.

3 Sustainability Maintenance

The beneficiaries should be considered in mind before developing the AI and IoT enabled solutions for the public realm. The experimental development deals with the testing of minimal viable product with the early adaptors before developing it in the lively passion [17]. Moreover, emission of certain toxic gases could also be reduced, thus providing an improvement to the environment [18].

A report estimates that the electric vehicle drivers of the trucks are more satisfied in their job by having a smooth ride and they appreciate in the non-emission of gasoline, thus encouraging a cutting-edge technology. This also encourages in recruiting new employees for transportation. Although some advantages overflow, certain drawbacks tend to reduce the efficiency. Normal commercial fuel vehicles could run more than 300 miles in a full tank. But it is impossible for EVs that could run only 100 miles when the fleet runs out of charge. Full recharge of the electric vehicle could last for 8 h [19].

Auditing, in common, is a disciplined assessment that involves analyzing, testing and affirmations of events and disciplines whose objective is to confirm if the organization obey with authorized necessities, interior policies, and established practices. The main purpose of the audit was to decide whether the Department of Energy has successfully as well as resourcefully reorganized its fleet vehicles to execute sustainability creativity at preferred venues [20]. Certain principles are followed by the sustainable fleet management that tries to reduce the environmental impacts by confirming the usage of cleaner vehicles, fuel-efficient process and driving as well as it sustains its task by controlling the traffic caused by the vehicles. In doing these activities, the fleet tries to reduce the usage of fuel and the cost of the vehicles and enhances the safety and benefit of workers. It also minimizes the troubles that could cause a blockage to the fleet [21]. These typical strategies could help us in maintaining the policies and certain conditions within the organization. For instance, efficiency of the business administration, safety as well as healthy, equal opportunities as well as enclosure, and community social responsibility (CSR) could be maintained. Moreover, the organization comes forward to report the amount of carbon emission from their vehicle during the time of transportation that could be reported in the form of Scope 3. The IoT-based sustainability fleet management could be followed based on three main strategies:

Demand management in the smart transport This particularly aims in the distance the fleet must travel. Moreover, the weather condition of the route could also be noted so that the responsible person could nominate a vehicle enough for the transportation, therefore reducing costs as well as effective delivery of goods. In fact for a shorter distance and less transportation goods size of the fleet could be reduced, which involves in saving the fuel effectively [22].

Sufficient technology and fuel usage This particularly aims in reporting the lowest emitting fuel vehicle. Vehicle which is highly maintained that could consume less fuel for miles and miles of distance could be preferred. In this case, the electric fleet could be used that could be more economical and in fact usage of bio-fuel for this purpose is also appreciated [23].

Measuring the efficiency of the fleet This aims in testing the effectiveness of the vehicle. The key role players of this domain are the drivers who are in-charge of that fleet. They should maintain the fleet that involves in cutting the fuel emissions and running costs. Certain principles could not be followed as per the desired requirement. But when an audit enters into a fleet industry, these reports should be necessarily maintained [24]. The key role played by the sustainability audit manager in certain criteria is shown in Fig. 2 given below.

IoT-controlled environmental auditing is necessarily a management tool for the environment that could rely on purely monitoring the environment with the certain basis of standards and criteria. Based on the classification of the standard and

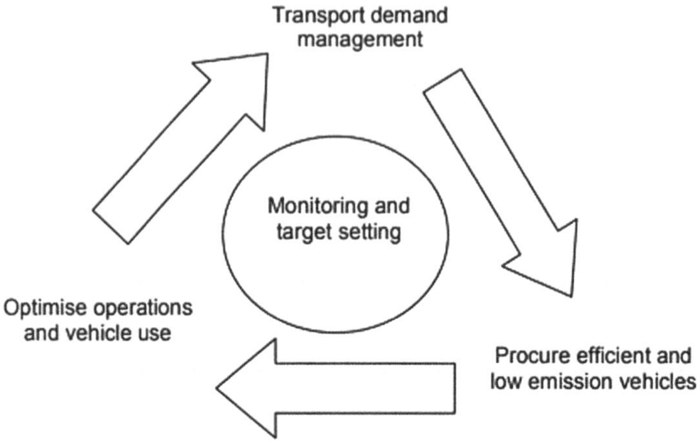

Fig. 2 Key areas concentrated by the IoT-based sustainability transport

the audit's focus area, they can be classified widely [25]. Every industries and organizations have understood the real necessity of environmental impact and they understood that the report could be checked out by various concerned parties.

The key role played by an IoT-controlled environmental audit involves the following key factors:

- Investigation
- Understanding
- Identification

With the help of these reports, the existing activity performed by the person or an individual could be reduced with the report generated and this could be very helpful for increasing the ambience of the nature. The audit is responsible for making a clear documentation and report in a systematic manner regarding the impacts caused by the fleet due to the environment. The reasons due to the audit are that people could understand the issue and also legislation could take necessary action when the happening goes out of the control [26].

Both the audits work for the welfare of the environment. Only certain variations could be found in both. The documentation done by both the audit could be publicly revolved. Sustainability audit plays an important role in the maintenance of fleet. They play different role in analyzing the fleet that could be sent for different transportation purposes with their complete documentation [27]. But this is not bothered regarding an IoT-controlled environmental audit who adds just an impact of the environment by the fleet. But travelling 200 km a day in the electrical vehicle could be quite difficult since at times we could run out of charge. Popular Mahindra Reva e20 could travel at 120 km per single charge. EVA created through TUM create project, a collaboration between Germany and Singapore, could travel 200 km with the operation of the air conditioner. Let's consider the commons that could be found in using the electrical vehicles.

At the year of 2009, the UK government launches a new rule of using electric vehicles and other vehicles that greatly supports in the decarbonization process. High quantity of emission of substance, for example, carbon monoxide that has been in a high rate of about 1.0 g/km for petrol and 0.5 g/km for diesel could cause severe health hazard. Moreover, the loss that could be caused was about 85 £/ton. According to the estimation, this results in the heavy economical loss. The Commons BEIS (business, energy & industrial strategy) committee has declared that the committee involves in launching an investigation into electric vehicles (EVs). This is being operated according to the norms of the government that tries to phase out the petrol and diesel vehicle within the year of 2040 [28].

Meanwhile the external costs that could involve in this decision involve the degradation of certain natural resource that could result in the production of electricity. Company must be aware of certain criteria that could help them in maintaining the financial savings and it could improve the quality of the organization. It has been estimated that £19.50 will be the total cost for the expenditure of electricity, where £84 and £70 are for the consumption of petrol and diesel for travelling 1000 km. This estimation seems to be extremely cheaper when using electric vehicle for travelling. We use batteries for the storage capacity, and this could be reported in terms of kilowatt/hours. Generally commercial EVs are provided for a steady driving of about 3–6 h. This timing could change depending on the driving. Batteries are generally packed which consists of individual cells. This could include several wirings and certain electronic tools. These batteries vary in their shape based on the power consumption. This could be an external cost making factor. Another one to be considered is that typical charging of the vehicle needs time for electric vehicle where else this is not required for either petrol or diesel type vehicles that involve just filling the tanks [29].

These the factors that must be considered in using IoT-based electric vehicles.

3.1 Cutting of Coal and Natural Gas

Electric vehicles generally depend on the coal and natural gas. When the government of UK tries to reduce the usage of coal and gas for electricity production, then that could be a greatest impact. However, the next upcoming list of fuel is the nuclear source. This could be levelled with the nuclear resources. According to the estimation of the year 2015, 2016, and 2017, usage of nuclear fuel for power production has been in the same state. Nuclear energy could produce a greater amount of electricity when compared to the coal and natural gas. Moreover, solar and wind energy for power production could also be encouraged in a highest rate [30].

3.2 Principle of Multi-criteria Decision Analysis

The MCDA is a new method of decision-making and analysis purpose for selection of project that exerts difficult socioeconomic benefits. This could also be stated as multi-criteria decision-making that concerns with the structuring and decision solving as well as planning problem that leads to multiple decisions. In this analysis, the individual judgement could be taken individually, and it could be evaluated in the fitting dimensions. These criteria must be prejudiced between each other since they are given with the equal priority. Obtaining the weightiest decision could need maximum liability as well as proficiency from the decision-makers since the decision has substantial authority on the consequence of the assessment.

MCDA could deal with the extensive field of research operations and the developers of MCDA should understand it as being different in evaluation approach compared with the economics-based CBA. This analysis criterion has evolved as an idea at the year of 1988 since there continued a circumstance where market price played a key valve and the involvement of evaluation criteria was not under establishment. This involves in combination of two aspects that added market price with the effective resource allocation. But this idea did not work well since there was no proper plan made for the evaluation which resulted in the market failure. This problem was well satisfied and solved with the help of multi-criteria decision-making that made a theoretical evaluation. Their major concentration and consideration involve in the problem solving as well as finding out the problem causing factors. This is an engineering approach but not economic approach. Thus, methods proposed in MCDA/MCDM supported a preference structure showing favorites on the different criteria.

The following steps are to be considered in the MCDA:

- Determining the objective of decision-making process.
- Selecting appropriate parameters for the decision.
- Alternative selection.
- Representation of important weighing methods.
- Aggregation method.
- Result of aggregation-based decision-making.

3.3 Determining the Objective of Decision-Making Process

First the general statement should be given to the decision-making. Here our objective is to replace the petrol and diesel vehicles with the electrical vehicles.

3.4 Selecting Appropriate Parameters for the Decision

The decision that we chose should be consistent with other decision. Moreover, it should be extremely independent. It should fall under the same category and the decision selected should be feasible. The focus is that it should not mingle with the other alternatives [31].

3.5 Alternative Selection

When the criteria are not feasible, then an alternative should be taken as a backup element. This alternative should satisfy the following criteria:

- It should be readily possible.
- It should be a real one but not ideal.
- Possible alternative must be chosen.

3.6 Representation of Important Weighing Methods

The methods adapted under these weighing criteria could be either compensatory or outranking. The example that comes under compensatory involves Analytical Hierarchy Process (AHP), Fuzzy Multi-Criteria Decision-Making Process (FDM), etc. The outranking methods could include certain elimination process such as Elimination and Choice Expressing Reality (ELECTRE) and Preference Ranking Organization Method for Enrichment of Evaluations (PROMETHUS).

3.7 Aggregation Method

The process of aggregation will select the best alternative option from the obtainable options. According to a fleet management, it could be a product or an average or a function that highly involves in improving the alternative option [32].

4 An IoT-Based Solution to Enhance Intelligent Traffic Lights Management Employing Programmable Logic Controller

A programmable logic controller (PLC) plays a vital role in the field of micro-processor. It is very simple and easy for the users to handle since it could consist of hardware and software components which could handle all the industrial components, and the work of the engineer is to program the PLC using ladder logic which could do certain type of automation and control in the industrial equipment. The greatest advantage of using PLC is that they can be reprogrammed depending on the user who handles it [33]. They have an astounding effect on the automation since these PLC components are highly flexible and reliable at their implementation stage. Since the functionalities of the micro-controllers have been increased, their reduced cost had increased their scope and made it to be used in the variety of fields. PLC had been used in the field of hydraulics to control fluid flow in the place of electromechanical relays. In recent days, programmable logic controllers are established into an erudite and largely flexible component in the control system that can perform complicated mathematical calculations as well as work in a faster speed. Some leading PLC manufacturers are ABB, Allen Bradley, Honeywell, Siemens, Mitsubishi, Modicon, etc.

In the electropneumatic system, the output is said to be the movement of solenoid. This could be done either by using relays or with the help of PLC. The work of the relay is to get the input signal from the various sensors and perform the function of normally opened or normally closed operation. With the movement the solenoid is operated. This could also be done with the help of programmable logic controllers. The required logic diagram is made with the certain timers and sequential operation. At last the signal is provided to the solenoid to the final control element to carry out the operation at various cylinders.

Control system acts as the constituent division in this industrial world. The necessity of the control system is to automate the process without the need of the human intervention. This could avoid some human error during the process. This automation process could be greatly supported with the PLCs which could be used in the process of designing, testing, and optimizing the process. The basic advantage of automating a system: increase in the production, maximize the accuracy, promotes safety to the environment.

4.1 Electropneumatic Controllers

There are three main steps involved in the electropneumatic controllers: (1) Input devices, which could consist of various sensors and the signals obtained from them are transferred to the relays and controllers. (2) Signal processing devices that consist of set of relay switches or PLCs and finally (3) Output devices that activates the solenoid to make a mechanical movement and these output devices can be an alarm or some indicators. The electrical devices used in the controllers could be push buttons, limit switches, timers, relays, solenoids, temperature and pressure switches. A sensor such as proximity sensors which is used in the indication of the certain thing and various electrical counters are also used. This controller combines the electrical and pneumatic technology, which is largely used in much application.

Either AC or DC source signal is applied here. Compressed air is used in this process as a medium of working. The operating voltage is around 12–220 V.

4.2 *Programming a PLC*

Programming a PLC is not a complicated task when the enough programming technique is adopted. Many concepts can be used to enter a program in PLC: (1) Draw a ladder logic diagram, (2) Functional blocks, (3) Various Boolean expressions based on the low level, (4) High level programming language. But the most commonly used method is the ladder logic diagram which is an easy and efficient approach. This ladder logic could consist of programming logic in a way identical to the switching circuit. With the help of certain conventions, the ladder logic is converted into PLC ladder diagram. With the help of the cathode ray tube and the keyboard, we can integrate certain components into a logical ladder program. The input switches, relays are indicated in the form of a contact switches and the solenoids, relays, counters, and timers are denoted as coils in the PLCs. Each step of a ladder program is denoted as rung. The program must place the necessary contacts and coils depending on the process. This could be verified rung by rung and the result could be displayed in the CRT screen. The PLC ladder logic has two vertical lines which are also termed as rungs which are supplied by a positive voltage in the left side and zero voltage at the right side. The horizontal lines (rungs) are in the middle of these vertical rungs which are the process of the automation and the process could be carried out depending on the horizontal rung positions. Between these two sides are the horizontal rungs for the assumed power flow. Various operations are carried out in a PLC ladder diagram and they are arithmetic operations, timer operations, PLC bit logic operations, and comparison operation.

NO Contact of PLC: The representation of NO contact is to scan the input signal ON (1) in the specific bit address. When the signal is ON, then the switch tends to close and performs the operation. Shortly we could say that NO switch is opened normally and when the power flows it will act as a closed switch.

Contact of PLC: This is exactly the opposite of the NO switch, which will scan the input signal OFF (0) in the specific bit address. When there is no signal, then the switch tends to remain in the closed position which gets opened when the signal is turned ON.

4.2.1 Push Button

A push button plays a major role in starting and stopping an operation. This could be indicated by an open and closed switch in a PLC diagram. There is also manually operated push button which could be used for some emergency circumstances. The functionality of the push button is to move the position of the actuator into housing. This is supported by the spring mechanism which could open or close the contact. There are two types of push button, namely, momentary push button and maintained contact push button. Momentary push buttons will come to its actual position once the button is manually released and the maintained push button which is mechanically latched has a certain latching mechanism to hold its position. The push buttons can be categorized as normally opened (NO), normally closed (NC), and change over (CO) type. At the position of NO the switch is opened and during the actuator action the contacts are closed which could permit the flow of air. At NC position the air is forbidden in this position and during the actuator position the contacts are closed. Below Figs. 3 and 4 represent the position of the push buttons.

Limit Switches: Programming a PLC is not a complicated task when the enough programming technique is adopted. Many concepts can be used to enter a program in PLC: (1) Draw a ladder logic diagram, (2) Functional blocks, (3) Various Boolean expressions based on the low level, (4) High level programming language. But the most commonly used method is the ladder logic diagram which is an easy and efficient approach. This ladder logic could consist of programming logic in a way identical to the switching circuit. With the help of certain conventions, the ladder logic is converted into PLC ladder diagram. With the help of the cathode ray

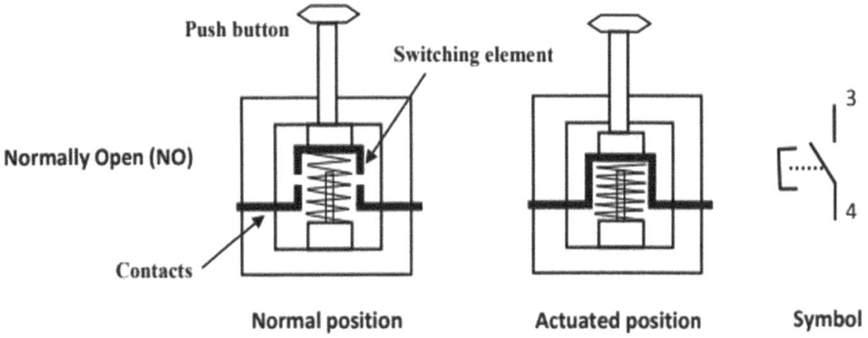

Fig. 3 Operation of push button when it is normally opened

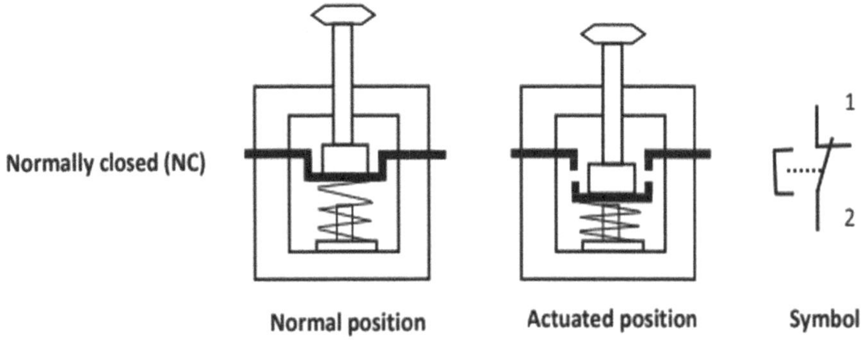

Fig. 4 Operation of push button when it is normally closed

tube and the keyboard, we can integrate certain components into a logical ladder program. The input switches, relays are indicated in the form of a contact switches and the solenoids, relays, counters, and timers are denoted as coils in the PLCs. Each step of a ladder program is denoted as rung. The program has to place the necessary contacts and coils depending on the process. This could be verified rung by rung and the result could be displayed in the CRT screen. The PLC ladder logic has two vertical lines which are also termed as rungs which are supplied by a positive voltage in the left side and zero voltage at the right side. The horizontal lines (rungs) are in the middle of these vertical rungs which are the process of the automation and the process could be carried out depending on the horizontal rung positions. Between these two sides are the horizontal rungs for the assumed power flow. Various operations are carried out in a PLC ladder diagram and they are arithmetic operations, timer operations, PLC bit logic operations, and comparison operation.

Limit switches are like the push button and the contradictory performance is that the push button is manually operated, and the limit switches are said to be mechanically actuated. The switch performs with the position of the fluid which could be a piston rod or a motor shaft. The limit switch provides an electric signal that provides an applicable system response. The limit switch is classified into two types based on the actuation of the contacts: (1) Lever-actuated contacts, in which the contacts operate at a very slow rate and (2) Spring-loaded contacts that make the contacts to perform expeditiously.

4.2.2 Pressure Switches

Pneumatic-electric signal conversion is done with the help of the pressure switch. The work of the pressure switch is to detect the transformation of the pressure change. With this information the electrical switch will be opened and closed when a prearranged pressure point is attained. Diaphragm or bellows will act as a pressure sensor. The change in the pressure is detected by the diaphragm which is in the form of a plate that contracts and expands. In this similar way, the bellows also react to pressure change. The pressure that comes through the inlet is noticed and when this pressure reaches a limit then the diaphragm or the bellows will expand which will make a spring-loaded plunger to break/make the contact.

4.2.3 Solenoids

In the electropneumatic controllers, the control valve which is electrically operated to actuate is the part which forms as an interaction layer. The important activity of the directional control valves is to regulate the supply of air (i.e., it should either switch to ON or OFF position), compression and rarefaction of the cylinder drives. This switching operation could be made possible with the help of the solenoid. The mode of operation is of two types: (1) Spring valves which return to its position until the power supply is provided to the solenoid. (2) Double solenoid valves cling to the finally operated state although when the solenoid is not supplied with the voltage. During the first step, every solenoid will not be energized and hence the directional control valve does not operate which make them inactive. There is no beginning position for the double valve since it will not return the spring to its position. The operating voltages of every solenoid are 12 V DC, 12 V AC, 12 V 50/60 Hz, 24 V 50/60 Hz, 110/120 V 50/60 Hz, 220/230 V 50/60 Hz. The numerous ways in which the solenoid operates are of the following: (1) 3/2 way single solenoid valve, spring return. (2) 5/2 way single solenoid valve, spring return. (3) 5/2 way single double solenoid valve.

4.2.4 Relays

This is a very simple device which could withstand any harsh environment and these relays are said to be electromagnetically operated. These relay switches act as a signal processing device. The relays are designed in such a manner to withstand the heavy power surges which could cause damage to the circuits. The system consists of the coil core through which the voltage is applied. This coil converts the electrical power into the electro-magnetic energy which will attract the armature towards the winding. The armature will actuate the relay, making it to be in the open or closed position. Depending upon this operation, the system provides the output. There is a spring attached in order to return the armature to the initial position. There is an interlocking capacity supported by the relays. This interlocking capacity will avoid the instantaneous switching of relays to ON and OFF. The relays are indicated by K1, K2, and K3. The relay contact with their operation is indicated in Fig. 5 given below.

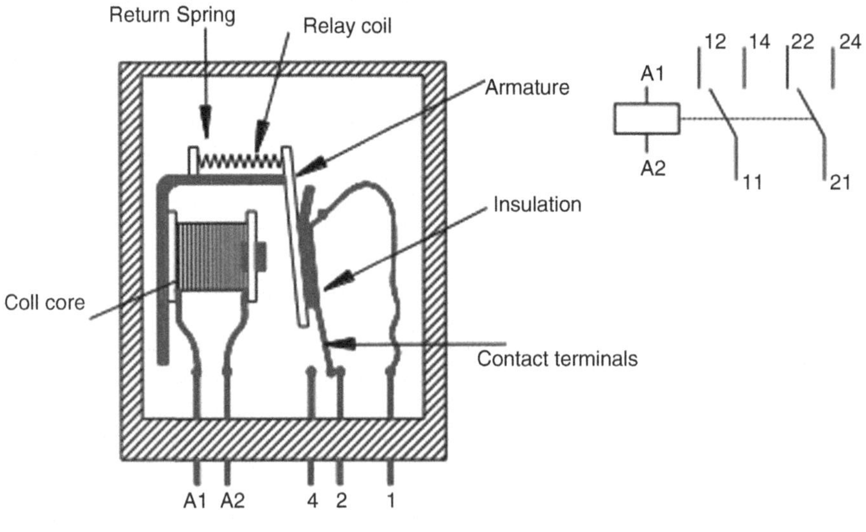

Fig. 5 Cross-sectional view of relay

5 IoT-Based Timers for Smart Traffic Lights Management

Timers for traffic lights play a major role in the smart transportation. In the pneumatic process when one side of the cylinder expands then the other side should contract, and this process is done successfully with the help of timers which could provide certain delay in the time of expansion and compression. Every control task which is automated needs a time to perform perfectly. These time delays find their own memory spaces in the system. The representation of the timer circuit purely depends on the manufacturer. The timer implemented in the PLCs is in the form of the software module and it could provide the digital way of representing the timing. The typical PLCs could have timers with the bit address of 64, 128, 256, 512, and many more. The timers in the ladder program could be indicated as T1, T2, etc. To explicitly reset timer, the logic of 1 should be applied to the reset port. The timers are of two types, namely, on-delay timer and off-delay timer. When the signal is received in the start input signal, the timer will be in the ON state. The output signal changes from 1 to 0, till the preset timing is attained (there will be a delay in the timing operation). When the signal reaches a preset timing, then the output will be changed from 0 to 1. The function of on-delay timer is described in Fig. 6 given below.

5.1 *Off-Delay Timer*

This is exactly the opposite of the on-delay timer. When the start input is provided with some signal, then the timer starts to operate and gives the output 1. When the result logic changes from 1 to 0, then the timer will not stop at a sudden. There is a delay in their off time till a preset time. When this time is reached, the output will change to 0. The function of off-delay timer is described in Fig. 7 given below.

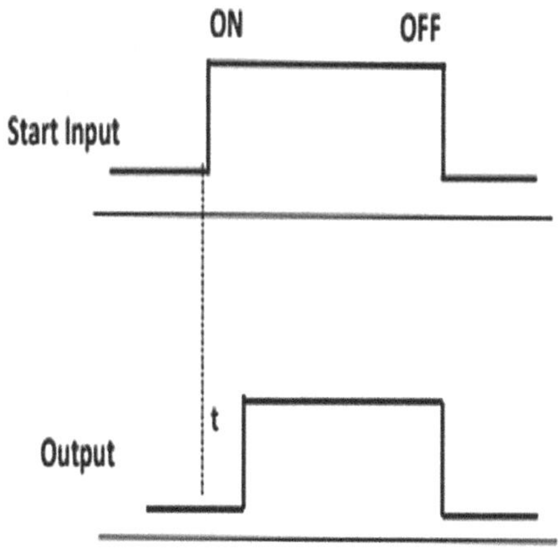

Fig. 6 Circuit diagram of on-delay timer with their timing variations

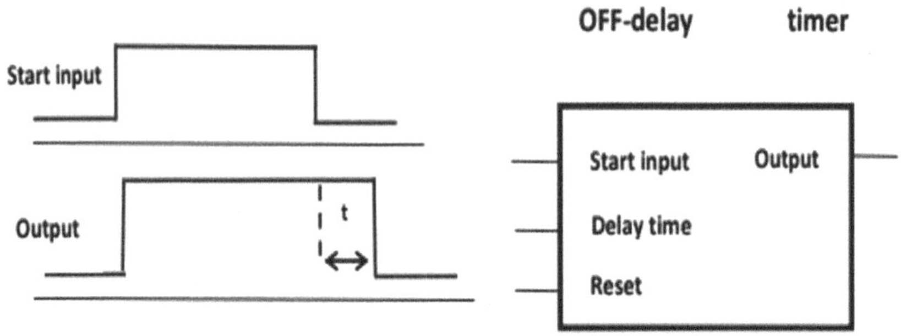

Fig. 7 Circuit diagram of off-delay timer with their timing variations

Fig. 8 Circuit diagram of up and down counter

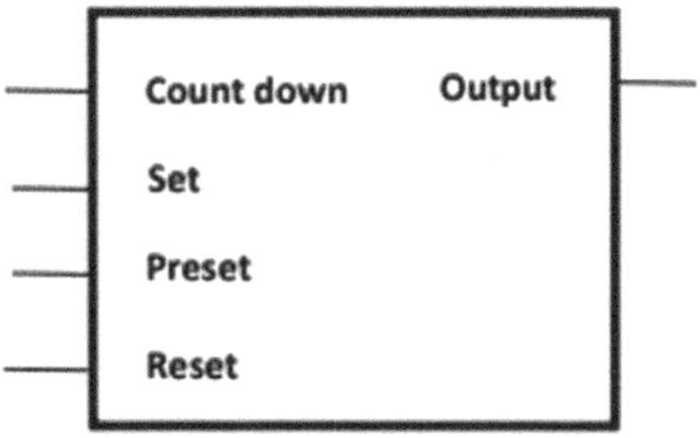

5.2 *Counters*

The count of the events and the particles kept to the process is indicated with the help of the counters. The controllers should be operated with the help of the counters. For instance, let's see the bottle filling mechanism. The counter counts the number of bottles which is to be filling with the solution. When the conveyor belt moves, the counter counts the bottles kept in the conveyor belt. This could be used in the place of a sorting device. There are two types of counters: (1) Up counter—in this counter, the accumulator value will be increased to 1 how long the input is set to true. (2) Down counter—this works by decrementing the accumulator value until a preset value is reached. The circuit diagram of up and down counter is shown in Fig. 8 given above.

5.3 *Memory Elements*

The transitional values are placed in the memory elements. They are represented as flags. The operation is performed by set and reset coil in the ladder diagram. Figure 9 shows the set and the reset coil. Let us denote the output as D. When the switch A gets closed, then the latch sets itself to logic 1 and the output will be 1.

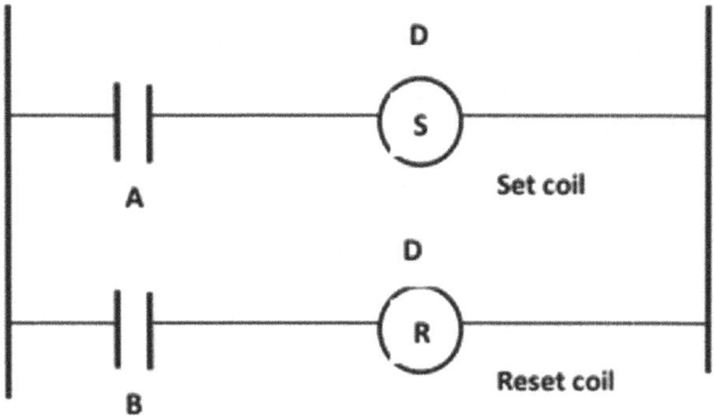

Fig. 9 Ladder logic diagram to indicate the latching function

This output remains until the switch B performs its operation. When the switch is opened, then also latching operation is continued. When the switch B gets opened, then unlatching process takes place which will set the output D as 0.

5.4 Replacement of Equipment with the Ladder Diagram

Relays are said to be an electro-magnetic switch and here we'll get into a small example of wiring a relay circuit into a ladder diagram. This is very necessary since the PLC does not understand code. Many PLCs convert the ladder diagram into coding. The following are the steps to be carried out.

Step 1: The PLC does not care about the external equipment and devices. These devices should be indicated by ladder symbol. For instance, the input devices are not cared by the PLC. Either it analyzes whether the given input is ON or OFF and the process it does is considered.

Step 2: The input voltage is very necessary. The AC supply is not considered. Rather the voltage is considered with is indicated in the form of two vertical bars (rungs) for the positive voltage and the ground.

Step 3: This could be indicated in the form of a circular symbol. This could be a bell or an alarm or a motor.

Step 4: We must indicate whether the PLC is loaded (which means addressing a device). Then the process is done rung by rung in a horizontal manner. These could be seen in the modicon equipment. The program does its operation and we can set the switch to automatic or manually operated depending on the user who handles the equipment. This is purely based on the engineer who handles it. He should have the entire knowledge regarding the PLC.

Step 5: Other than this we have temperature switch which could sense the temperature. These switches can be wired either normally opened or normally closed depending on the temperature variations. Other than these, sensor plays a major role in the field of PLC.

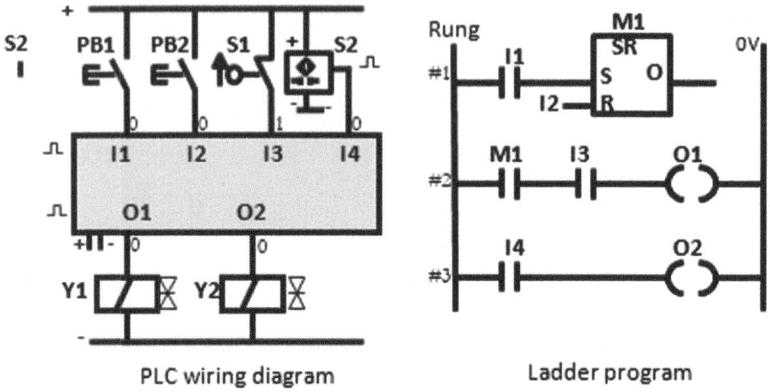

Fig. 10 PLC wiring with the ladder logic of a pneumatic controller

The pneumatic controller with the implementation of the smart traffic is quite easy since it reduces the size, is very easy to maintain, user friendly, reduce costs burden, easily programmable as shown in Fig. 10. Although we have some difficulties in finding the error and wiring becomes a challenging task, this is very rugged to certain temperature and vibrations. Smart traffic management and monitoring of congestion has become very important in many countries and has spawned the need for management schemes such as traffic lights and efficient land use. With regard to traffic lights, there are mainly two strategies that are used: fixed and real time. The former works with preset timing, and the duration of the signal displayed on the light is preset from experience giving no consideration to the current status. The latter relies on real-time conditions of the road; decisions of how to manage traffic lights are based on the prevailing traffic parameters such as count of cars, their speeds and direction.

6 Conclusion

Smart way transport recognizes the operators of the transport system and shippers who help them in improving their environmental facilities by certain measures. Those operators should involve in cost savings and fuel efficiency to improve the smart urban planning strategy once in 3 years. Through the process of information exchange, certain intelligent functions could be achieved by IoT through the process of detecting, tracking, learning, communicating, connecting, tracing, monitoring, searching, evaluating, controlling, operating, planning, repairing, and managing things.

References

1. Dasgupta, A., Nagaraj, R., & Nagamani, K. (2016). An internet of things platform with Google Eddystone beacons. *Journal of Software Engineering and Applications, 9*(6), 291–295.
2. Dijkman, R., Sprenkels, B., Peeters, T., & Janssen, A. (2015). Business models for the internet of things. *International Journal of Information Management, 35*(6), 672–678.
3. De Haan, F. J., Rogers, B. C., Frantzeskaki, N., & Brown, R. R. (2015). Editorial. Transitions through a lens of urban water. *Environmental Innovation and Societal Transitions, 15*, 1–10.
4. Domènech, L., March, H., Vallès, M., & Saurí, D. (2015). Learning processes during regime shifts: Empirical evidence from the diffusion of greywater recycling in Spain. *Environmental Innovation and Societal Transitions, 15*(1), 26–41.
5. Udendhran, R. A hybrid approach to enhance data security in cloud storage. In *ICC '17 Proceedings of the Second International Conference on Internet of things and Cloud Computing* at Cambridge University, UK—March 22–23, 2017. ISBN: 978-1-4503-4774-7. https://doi.org/10.1145/3018.
6. Suresh, A., Udendhran, R., Balamurgan, M., et al. (2019). A novel internet of things framework integrated with real time monitoring for intelligent healthcare environment. *Journal of Medical System, 43*, 165. https://doi.org/10.1007/s10916-019-1302-9.
7. Suresh, A., Udendhran, R., & Balamurgan, M. (2019). Hybridized neural network and decision tree based classifier for prognostic decision making in breast cancers. *Journal of Soft Computing*. https://doi.org/10.1007/s00500-019-04066-4.
8. Archer, J., Fotheringham, N., Symmons, M., & Corben, B. (2008). *The impact of lowered speed limits in urban and metropolitan areas*. Melbourne: Monash University Accident Research Centre.
9. Baas, P., & Latto, D. (2007). *Heavy vehicle efficiency*. Auckland: Transport Engineering Research New Zealand.
10. Baas, P., Latto, D., & Ludvigson, T. (2007). *Light vehicle fleet—energy use*. Auckland: TERNZ: Vehicle Management.
11. Baas, P. (2008). *Fleet commitment initial work programme trial*. Auckland: TERNZ: Vehicle Management.
12. Baas, P., & Taramoeroa, N. (2008). *Analysis of the safety benefits of heavy vehicle accreditation schemes*. Sydney: Austroads.
13. Bomel Ltd. (2004). *Safety culture and work-related road accidents*. London: Department for Transport.
14. Charlton, S., Mackie, H., Baas, P., Hay, K., Menezes, M., & Dixon, C. (2010). Using endemic road features to create self-explaining roads and reduce vehicle speeds. *Accident Analysis and Prevention, 42*, 1989–1998.
15. de Pont, J. (2010). *An overview of commercial vehicle operations in New Zealand*. TERNZ: Auckland. 66pp.
16. Department for Transport (DfT). (2004). *Review of the road haulage modernisation fund*. London: DfT.
17. DfT. (2011). *Freight best practice*. Retrieved from http://webarchive.nationalarchives.gov.uk/+/http:/www.freightbestpractice.org.uk/
18. Energy Efficiency and Conservation Authority (EECA). (2011). *Energy end use database*. Retrieved from http://www.eeca.govt.nz/energy-end-use-database
19. Energy Efficiency and Conservation Authority (EECA). (2011). *The energy spot*. Retrieved from http://www.eeca.govt.nz/node/6249
20. Government Accountability Office. (2012). *BPR glossary of terms*. Retrieved February 15, 2012, from http://www.gao.gov/special.pubs/bprag/bprgloss.htm
21. Hattie, J., & Timperley, H. (2007). The power of feedback. *Review of Education Research, 77*(1), 81–112.

22. Joint Transport Research Centre. (2012). *Transport outlook 2008, focussing on CO2 emissions from road vehicles*. Retrieved from http://www.internationaltransportforum.org/jtrc/DiscussionPapers/DP200813.pdf
23. Issling, C., & Coyle, M. (2008). *Survey of key influencers, fleet operators and drivers of heavy and light commercial vehicles in New Zealand to ascertain the practices, attitudes, perceptions and barriers to change in relation to introducing fuel efficiency interventions*. Christchurch: Kissling Consulting Ltd and IMISE Ltd. 122pp.
24. Lawson, K., Michaelis, C., & Waldron, D. (2008). *SAFED impact assessment 2007 final report*. Birmingham: Databuild Research and Solutions. 37pp.
25. Maxwell, C. (2004). *Occupational health and safety act review*. Melbourne: State of Victoria. 415pp.
26. McKenna, F., Waylen, A. E., & Burkes, M. E. (2009). *Male and female drivers: How different are they?* Basingstoke: AA Foundation for Road Safety Research.
27. Ministry for the Environment. (2009). *New Zealand's fifth national communication under the United Nations framework convention on climate change*. Wellington: Ministry for the Environment.
28. Ministry of Economic Development. (2011). *New Zealand energy data file*. Retrieved from http://www.med.govt.nz/sectors-industries/energy/pdf-docs-library/energy-data-andmodelling/publications/energy-data-file
29. NZ Transport Agency. (2009). *Vehicle selection guide*. Retrieved from www.nzta.govt.nz/vehicle/choosing/heavy/index.html
30. Office of Energy Efficiency. (2007). *Smart driver program for heavy vehicles*. Ottawa: Office of Energy Efficiency.
31. Ministry of Transport (MoT). (2008). *Safe and fuel efficient driving tool kit*. Wellington: MoT. 20pp.
32. MoT. (2011). *Connecting New Zealand*. Wellington: MoT. 45pp.
33. Joseph, C., Kishoreraja, C., Reji, M., & Baskar, R. (2017). Design of wireless sensor network gateway for the internet of things. *International Journal of Pure and Applied Mathematics, 117*(9), 41–45.

Trust Based Data Reduction in Sensor Driven Smart Environment

N. Karthik and V. Ebenezer

1 Introduction

Our environs will slowly change into smart environment, where smart intelligent sensors and actuators are installed to execute their task autonomously. Smart environment includes effective monitoring and protecting the environment with the help of sensors. Wireless sensor network (WSN) is capable of collecting data from environment, processing it and taking decisions in real time [1]. Sensor nodes are often deployed in smart environment to assist monitoring and protecting the environment. Large quantity of data is brought from sensor nodes periodically for smart environment. One of the biggest issues in sensor network based application is resource restriction. Sensor nodes are equipped with small battery, small computing device and low power radio antenna for monitoring physical and environmental conditions. Uninterrupted transmission of generated data between source sensor nodes and sink node needs higher battery power. Energy needed for data communication is higher than the energy needed for processing and sensing. To decrease the resource consumption of sensor nodes, we can minimize the number of data communication between node and sink node [2]. Data reduction is one of the techniques used for minimizing the number of data communication and thus decreasing the resource utilization [3]. Data reduction process is assorted into three categories, namely in-network processing, data aggregation and data prediction.

The aim of in-network processing is to decrease the amount of data while travelling through intermediate nodes towards sink node [4]. Data aggregation is the procedure of summing up of data from similar nodes and transmitting aggregated

N. Karthik (✉) · V. Ebenezer
Department of Computer Science and Engineering, Karunya Institute of Technology and Sciences, Coimbatore, India
e-mail: ebenezerv@karunya.edu

G. R. Kanagachidambaresan et al. (eds.), *Internet of Things in Smart Technologies for Sustainable Urban Development*, EAI/Springer Innovations in Communication and Computing, https://doi.org/10.1007/978-3-030-34328-6_4

data to sink nodes thus reducing the number of data transmissions. Data prediction is the process of predicting the future upcoming data by constructing the prediction model from history of data. The model parameters are shared to sink node for dual data prediction scheme and thus reducing the number of data transmission. All three approaches of data reduction techniques focus on reducing the number of transmissions [2–21]. None of the approaches focuses on data quality, trustiness of data and intermediate node/aggregator before performing data reduction operation. Malicious attacks, trustiness of data, and node are considered in proposed trust based data reduction technique to increase the reliability of the application.

1.1 Contributions

The main contributions of this chapter are given as follows: (1) We constructed a trust based data reduction scheme for data prediction to ensure that only trustworthy nodes and data are considered for prediction process. Further, the trust based data reduction scheme is constructed to prolong the lifetime of application by avoiding the unwanted data transmissions. (2) We conducted extensive set of experiments with real sensor data to show that the absence of trust mechanism in data reduction process will reduce the lifetime of the application. (3) The results show that the proposed scheme treats the untrustworthy nodes, malicious nodes, untrustworthy data and missing data effectively.

1.2 Motivations

(1) Data transmission is widely believed as the dominating energy consumption factor in sensor nodes than sensing and data processing. Ninety-four percent of energy is consumed on an average for data transmission in standard sensor node [19]. (2) Three to sixty percent of real sensor data were found to be faulty in many real time sensor applications [18]. (3) Real time application suffers from notable data losses [17]. (4) The presence of untrustworthy nodes and untrustworthy data in data prediction and data reduction process decreases the reliability of the application which leads to wastage of resources.

1.3 Organization

Rest of the chapter is developed as follows: Sect. 2 explains about existing data reduction schemes in WSN. The proposed trust based data reduction scheme is introduced in Sect. 3. Section 4 talks about the details of experimentation and performance metrics. Analysis of outcomes is highlighted in Sect. 5. Section 6 closes our chapter with conclusion and future work.

2 Related Work

We talk about the recent works of data reduction schemes of WSN in this section. Data reduction in WSN is the procedure of cutting down the number of data transmission from source node to sink node [22]. The data is forwarded hop by hop from source to sink node through various intermediate nodes. Massive quantity of data is collected from periodical monitoring applications of WSN. In this scenario, sensor node needs to forward the data frequently to sink node for analysis. The amount of energy consumption is directly proportional to amount of data transferred to sink node. Around 94% of energy is utilized for data communication in WSN. Only 6% of total energy is used for sensing and data processing. Because of redundant deployment of nodes in the networks, there will be redundant data generation. Transmission of redundant data to sink node leads to energy wastefulness in the network and it decreases the lifetime of the application. Therefore, data reduction scheme must be designed to decrease the number of data communication and thus reducing the energy expenditure. Data reduction schemes can be categorized into three types: in-network processing, data aggregation and data prediction as shown in Fig. 1. In in-network processing, quality of node and data is checked in source and neighbour nodes before data transmission to facilitate distributed data fault detection techniques. If any erroneous data or untrustworthy node is found, then it is ignored by in-network processing schemes to save the energy dissipation.

Data aggregation is the procedure of removing duplicate data from WSN to conserve the energy. Due to faulty node and faulty data, the performance of data aggregation is decreased. And also it takes more energy for data retransmission in case of any data faults and malicious node. So data aggregation scheme must be designed to reduce the effect of faulty data, untrustworthy nodes and unreliable link. The third category of data reduction in WSN is data prediction. In this

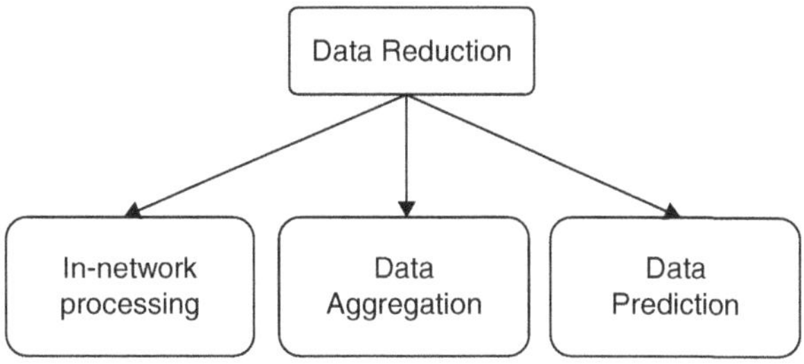

Fig. 1 Categories of data reduction schemes

process, the data is predicted at source node and sink node to reduce the data transmissions. The model is generated at source node based on historical data and parameters of model will be shared to sink node for data prediction. Without the transmission of actual data, the model is used to predict the sensor data at sink node and thus reduces the energy consumption. Due to the presence of faulty data in historical data, the parameters generated for data prediction are not reliable. In existing techniques of data aggregation, faulty data, untrustworthy nodes and unreliable links are not considered. In data prediction models, faulty data, data of untrustworthy nodes are considered for generation of model parameters and prediction. In in-network processing, spatio-temporal attribute correlation is considered in existing techniques. However, the data density correlated trustworthy node is not considered for identifying data faults and untrustworthy nodes. Recent works of data aggregation with trust mechanism for data reduction are shown in Table 1. Most of existing schemes of data aggregation focus on energy efficiency, delay awareness. None of them considers trustiness of link, node and data together for achieving energy efficiency, delay aware trustworthy data aggregation for data reduction. Recent works of trust based in-network data processing for data reduction are shown in Table 2. Most of existing schemes of in-network data processing focus on correlation, provenance based data reduction. None of them considers data density, provenance and correlation together for data fault detection and data reduction.

Recent works of correlation based data reduction in WSN are shown in Table 3. Most of existing schemes of reduction focus on temporal, spatial and attribute correlations. None of them considers node trustiness and data trustiness for data reduction.

Table 1 Recent works of trust based data aggregation

Ref	Year	Energy efficiency	Delay aware	Node trust	Data trust	Link trust
12	2016	Yes	No	No	No	No
13	2016	No	No	No	No	No
14	2016	Yes	No	No	No	No
15	2018	Yes	No	No	No	No
16	2018	No	No	No	Yes	No
21	2019	Yes	Yes	Yes	Yes	No

Table 2 Recent works of in-network data processing

Ref	Year	Temporal correlation	Spatial correlation	Attribute correlation	Provenance based trust	Data density
17	2016	Yes	Yes	No	Yes	No
18	2017	Yes	Yes	Yes	No	No
19	2017	Yes	Yes	Yes	Yes	No
20	2017	Yes	Yes	Yes	No	No
21	2019	Yes	Yes	Yes	No	Yes

Table 3 Recent works of data reduction

Ref	Year	Temporal correlation	Spatial correlation	Attribute correlation	Node trustiness	Data trustiness
6	2011	Yes	Yes	Yes	No	No
7	2017	Yes	Yes	Yes	No	No
9	2017	Yes	Yes	No	No	No
4	2018	Yes	Yes	No	No	No
8	2019	Yes	Yes	No	No	No

3 Proposed Trust Based Data Reduction Scheme

In this section, a trust based data reduction scheme is proposed as shown in Fig. 2.

The source node uses temporal correlation metric for in-networking data processing and data prediction. The upcoming data is predicted by using time series based temporal correlation coefficient. The quality of predicted data is based on the trustiness of node and trustiness of history of data. Application domain experts define restricted interval and threshold limits for every phenomenon. The predicted data is compared with restricted interval and threshold limit to find the quality of data. The trust value lies between -1 and $+1$ [16]. The temporal coefficient based trustiness of data TCT is calculated by using Eq. (1). X_i represents data item of a node and PX_i represents predicted data item of the prediction model. The model is constructed with the help of parameters like correlation coefficients, trust value of nodes and data. The parameters are forwarded to sink node for data prediction. The neighbour node uses data density and spatial correlation for in-network data processing and data prediction. Data is predicted with the help of correlation coefficients and trust value of data and node. Instead of transmitting the entire data, the parameters include spatio-temporal coefficients, data density correlation coefficients are sent to sink node for data reduction. The spatial correlation SC among two closely located nodes is calculated by Eq. (2) where X_i represents the data item of a node and Y_i represents the data item of neighbour node. The spatial correlation based trustiness of data SCT is calculated by using Eq. (3). Multi-attribute correlation coefficient (MAC) is calculated for in-network processing and data prediction at aggregator node. MAC is calculated by using Eq. (4) where r represents the correlation coefficient. The MACT based trustiness of data is calculated by using Eq. (5). Node trustiness in terms of trustworthy data generation is calculated by using Eq. (6). Trustiness of node is defined as the ratio of number of trustworthy data generated to the total number of data generated. Resource based trustiness of node is calculated by using Eq. (7) where E represents the remaining energy of the node and $E\text{th}$ represents the threshold value of energy fixed by domain experts. Here energy is considered as the resource of a sensor node. Communication based trustiness of node is calculated by using Eq. (8). Total trust of a node is calculated by using Eq. (9). Node trust (NT) is calculated by using Eq. (10). Let us assume two data items Da, Db from two closely located nodes and

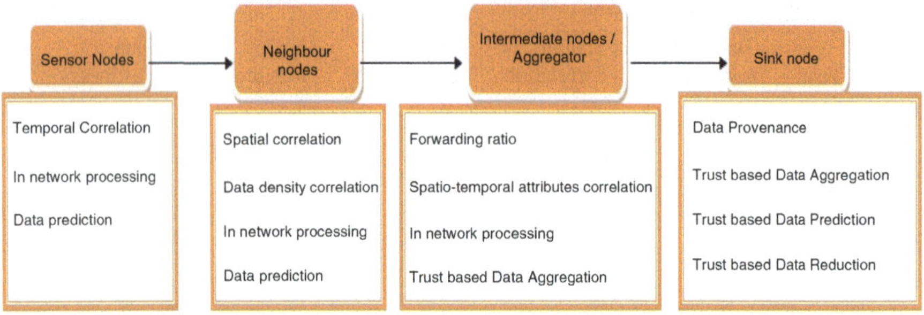

Fig. 2 Proposed trust based data reduction scheme

their provenance be Pa and Pb, respectively. Then data provenance based data trust (DPT) is calculated by using Eq. (11). The total data trust (DT) is calculated by using Eq. (12). Forwarding ratio (FR) is used to find the selfishness of the intermediate nodes. FR is calculated by using Eq. (13).

$$\mathrm{TCT} = \begin{cases} -1, \text{ if } |Xi - PXi| > \text{TCth, data fault} \\ 0, \text{ if } |Xi - PXi| = \text{TCth, uncertain} \\ +1, \text{ if } |Xi - PXi| < \text{TCth, trustworthy data} \end{cases} \tag{1}$$

$$\mathrm{SC} = \frac{\sum_{i=1}^{n}(Xi - \bar{X})(Yi - \bar{Y})}{\sqrt{\sum_{i=1}^{n}(Xi - \bar{X})^2(Yi - \bar{Y})^2}} \tag{2}$$

$$\mathrm{SCT} = \begin{cases} -1, \text{ if } |Xi - PXi| > \text{SCth, data fault} \\ 0, \text{ if } |Xi - PXi| = \text{SCth, uncertain} \\ +1, \text{ if } |Xi - PXi| < \text{SCth, trustworthy data} \end{cases} \tag{3}$$

$$\mathrm{MAC} = \frac{\sqrt{r_{xz}^2 + r_{yz}^2 - 2r_{xz}r_{yz}r_{xy}}}{1 - r_{xy}^2} \tag{4}$$

$$\mathrm{MACT} = \begin{cases} -1, \text{ if } |Xi - PXi| > \text{MACth, data fault} \\ 0, \text{ if } |Xi - PXi| = \text{MACth, uncertain} \\ +1, \text{ if } |Xi - PXi| < \text{MACth, trustworthy data} \end{cases} \tag{5}$$

$$\mathrm{NT1} = \frac{\text{(number of trustworthy data)}}{\text{(total number of data generated)}} \tag{6}$$

$$\mathrm{NT2} = \begin{cases} \text{Untrustworthy node, if} E < E\text{th} \\ \text{normal node, otherwise} \end{cases} \tag{7}$$

$$\mathrm{NT3} = \frac{(\text{successful transactions} - \text{ unsuccessful transactions})}{(\text{total number of transactions})} \tag{8}$$

$$\mathrm{TT} = w1\mathrm{NT1} + w2\mathrm{NT2} + w3\mathrm{NT3} \tag{9}$$

where $w1 + w2 + w3 = 1$

$$\mathrm{NT} = \begin{cases} -1, \text{ if TT} < -0.3, \text{ untrustworthy node} \\ 0, \text{ if } -0.3 < \text{TT} < 0.3, \text{ uncertain} \\ +1, \text{ if TT} > 0.3, \text{ trustworthy node} \end{cases} \tag{10}$$

$$\mathrm{DPT} = \begin{cases} -1, \text{ if Da!} = \text{Db\&Pa} = \text{Pb, untrustworthy data} \\ 0, \text{ if Da} = \text{Db\&Pa} = \text{Pb, uncertain} \\ +1, \text{ if Da} = \text{Db\&Pa!} = \text{Pb, trustworthy data} \end{cases} \tag{11}$$

$$\mathrm{DT} = w1\mathrm{TCT} + w2\mathrm{SCT} + w3\mathrm{MCT} + w4\mathrm{DPT} \tag{12}$$

where $w1 + w2 + w3 + w4 = 1$

$$\mathrm{FR} = \frac{(\text{number of data packets forwarded})}{(\text{total number of data packets supposed to be forwarded})} \tag{13}$$

4 Experimental Setup

We introduce our experimental setup, simulation parameters and performance metrics in this section. MATLAB is used to implement TDR. We used INTEL Berkley lab [23] setup with 54 sensor nodes deployed to gather temperature, light and humidity as depicted in Fig. 3. We assumed nodes 31, 32, 33 and 34 as normal nodes. Node 1 is considered as collector/aggregator, whereas node 3 is acting as a relay node. Node 6 is the sink node which performs data analysis and event detection process. Trusted AODV protocol is used for simulations. We used 5000 samples of multivariate data for experiment. The sampling time is 31 s. Data faults are inserted randomly apart from original data faults present in the dataset. Malicious behaviours like on and off, DOS attacks and replication attacks are simulated for experiments. Total energy is calculated from energy required for listen, receive, transmit and compute operation. Root mean square error (RMSE) is the difference between actual value and predicted value. We considered periodic monitoring application where the nodes involve in same number of data transmissions.

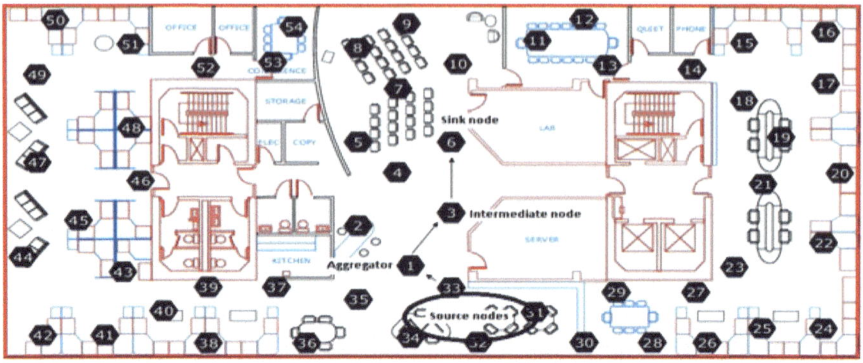

Fig. 3 INTEL lab sensor deployment

5 Results and Discussions

The analysis of energy consumption for data aggregation without trust process and trust based data aggregation is explained in this subsection. The remaining energy of a node plays an important role in the lifetime of the network. The probability of erroneous data generation is more when the nodes battery is draining out. Here we consider the data from nodes 31, 32, 33 and 34 for data aggregation. Node 1 is responsible for data aggregation process. The energy consumption for data aggregation includes the energy consumed for computation, listening, transmission and reception. The energy consumption for trust based data aggregation process includes the energy consumed for computation, listening, transmission, reception and trust computation. Figure 4 depicts the energy consumption analysis between data aggregation without trust and with trust process. Due to faulty data, retransmission of data required for traditional data aggregation, which leads to high energy consumption than trust based data aggregation, where data is predicted with trust based data prediction.

Figure 5 depicts the energy consumption analysis between data aggregation and trust based data aggregation by changing the number of untrustworthy nodes. Untrustworthy nodes exhibit selfish behaviour, untrustworthy data generation and suffer from resource dissipation. We varied the number of untrustworthy nodes to see its effects on data aggregation and trust based data aggregation. The traditional data aggregation process consumes more energy than trust based data aggregation process as shown in Fig. 5. The data aggregation process requires retransmission of faulty data and selection of trustworthy nodes for handling the untrustworthy nodes in the network. This is the main reason for high energy consumption than trust based data aggregation.

In this subsection, we test the operation of data prediction with and without trust process as shown in Fig. 6. INTEL lab dataset is used for data prediction. In addition to existing data faults, we randomly inserted data faults to check the performance of data prediction process. RMSE is used as performance metric. It is the difference

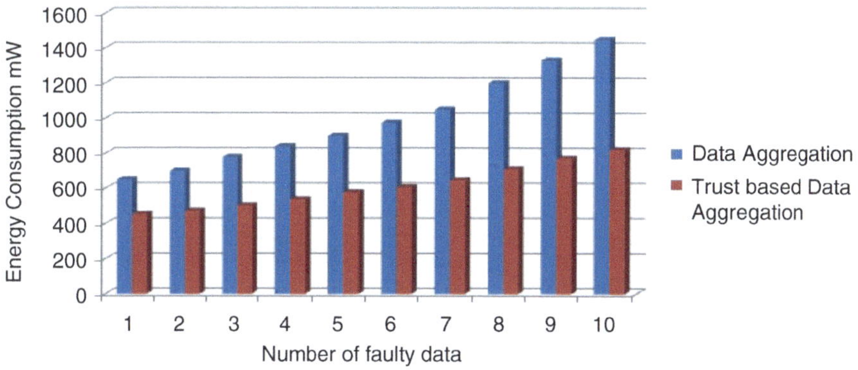

Fig. 4 Energy consumption analysis between DA and TDA for faulty data

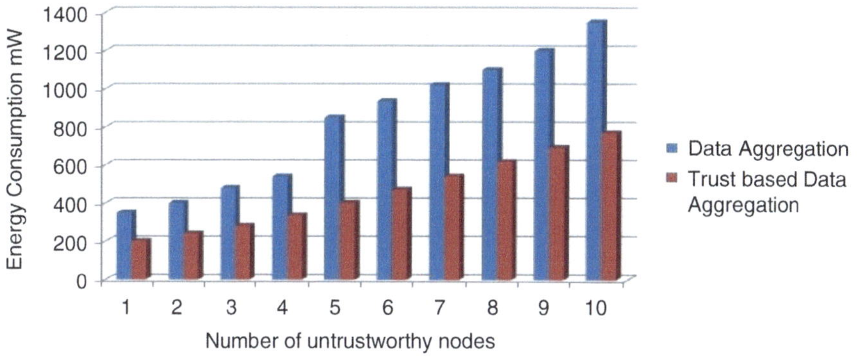

Fig. 5 Energy consumption analysis between DA and TDA for untrustworthy nodes

between actual and predicted value. The traditional data prediction process uses temporal, spatial and attribute correlation coefficient for data prediction. Trust based data prediction uses correlation coefficient, data provenance, trustiness of node and data for data prediction. When we vary the number of data faults from 1 to 11 with 10% increment in every sampling, result shows that the trust based data prediction achieves better error rate than data prediction without trust.

The proposed TDR is compared with trust based data gathering (TDG) [20] for detecting untrustworthy nodes as shown in Fig. 7. TDG is the recent trust based data aggregation method. The malicious attacks like DoS, on–off, replication attacks are simulated and we varied the untrustworthy nodes from 1 to 10 with an increment of 10% for every sampling in the experiment. TDG uses node trust, data trust, data density correlation, resource trust, data prediction and data provenance for detecting malicious nodes. Results prove that the proposed method outperforms recent TDG because it uses trust based data prediction and data provenance methods for detecting untrustworthy nodes.

Figure 8 shows the performance of detecting untrustworthy data by proposed TDR and existing recent method TDG. We randomly inserted the data faults in

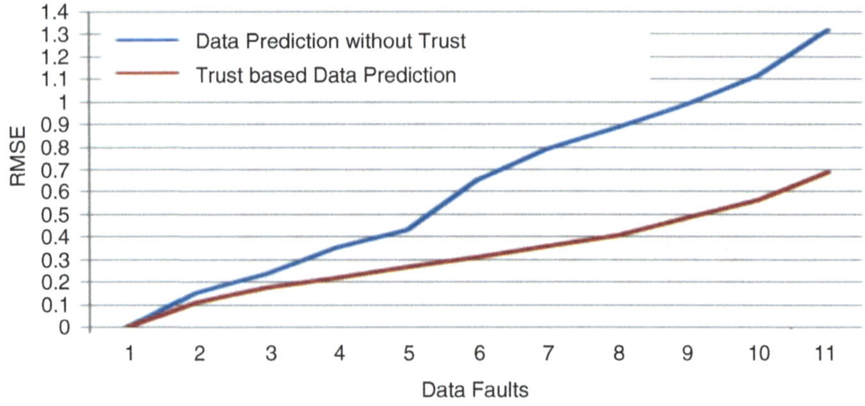

Fig. 6 Data prediction analysis with and without trust process

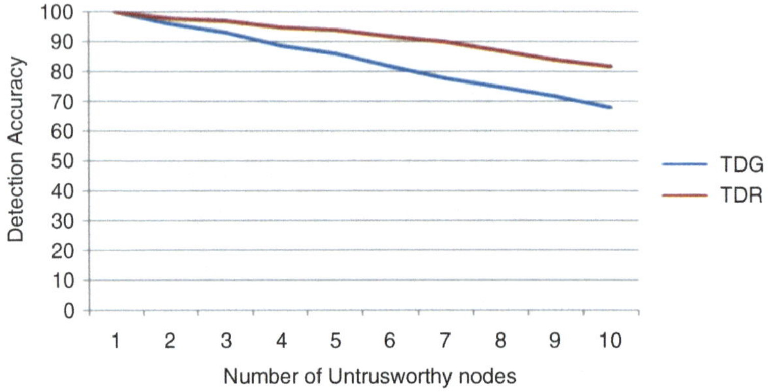

Fig. 7 Analysis of detection accuracy of untrustworthy nodes

existing dataset. We changed the number of data faults from 1 to 10 with an increment of 10% in every sampling. TDG uses data density correlation, node trust, data trust for detecting untrustworthy data faults, whereas proposed method detects the data faults with the help of node trust, data provenance and trust based data prediction. When we add the number of data faults in experiment, there is performance degradation in detection accuracy for both methods. However the proposed outperforms the existing TDG in detecting the untrustworthy data.

The performance of proposed TDR, data reduction without trust and traditional method is compared in number of data transmissions and depicted in Fig. 9. We considered eight (8) data transmissions per node for every sampling in spatio-temporal correlated region. We changed the number of normal nodes from 1 to 10 with an increment of 10% in every sampling period. All generated data packets are transmitted to sink nodes in traditional method. Data reduction without trust method uses data prediction to decrease the number of data transmission. However, the untrustworthy data, nodes, malicious attacks are considered for data reduction

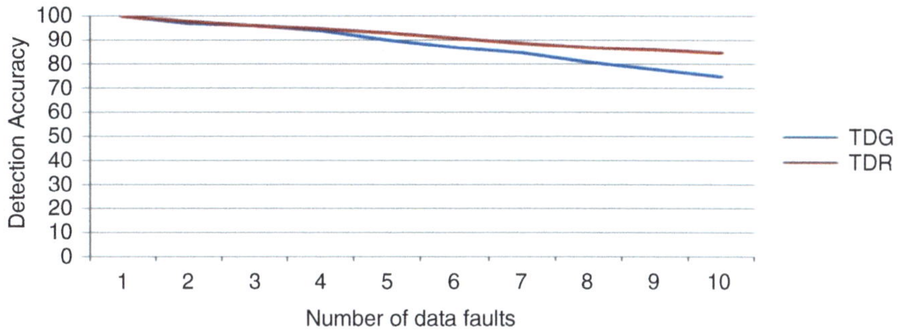

Fig. 8 Analysis of detection accuracy of data faults

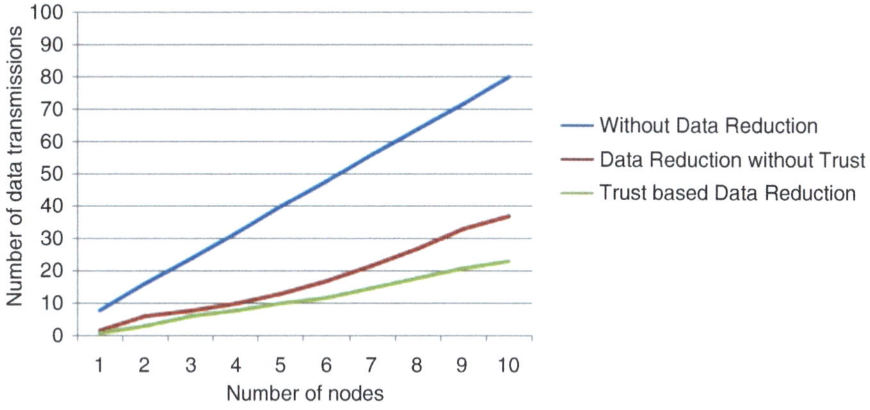

Fig. 9 Comparison of number of transmissions

without trust. In proposed method, trust based data prediction with node trust, data trust, resource trust, data provenance is considered for data reduction. By transmitting the model parameters, the data are predicted at sink node with same quality. Result shows that the proposed method achieves lesser number of data transmissions than traditional method and data reduction without trust method.

6 Conclusion

In this chapter, we introduced trust based data reduction scheme in sensor driven smart environment for data prediction, reducing unwanted data transmission and energy dissipation. INTEL lab sensor data is used to show that the proposed scheme can replace the data faults, data losses and data from untrustworthy nodes with the aid of data prediction with trust process and ignoring the transmissions of redundant and erroneous data for achieving energy efficiency and prolonging the lifespan of the application. We also showed that the unusability of trust mechanism in data prediction, data aggregation and data transmission process in sensor driven smart environment could affect the regular operation of the application. We are concentrating on data semantics in our next work for combining the data of different applications in trust evaluation to improve event detection accuracy.

References

1. Akyildiz, I. F., Su, W., Sankarasubramaniam, Y., & Cayirci, E. (2002). Wireless sensor networks: A survey. *Computer Networks, 38*(4), 393–422.
2. Vy, V. D., An, T. H., Nghe, V. D., Tu, B. D., & Du, H. Q. (2012, September). Data reduction algorithms for wireless sensor networks in environment monitoring and warning applications. In *2012 Fourth International Conference on Computational Intelligence, Modelling and Simulation* (pp. 416–421). Piscataway, NJ: IEEE.
3. Tan, L., & Wu, M. (2015). Data reduction in wireless sensor networks: A hierarchical LMS prediction approach. *IEEE Sensors Journal, 16*(6), 1708–1715.
4. Adawy, M. I., Nor, S. A., & Mahmuddin, M. (2018). Data redundancy reduction in wireless sensor network. *Journal of Telecommunication, Electronic and Computer Engineering, 10*(1–11), 1–6.
5. Arunraja, M., Malathi, V., & Sakthivel, E. (2015). Energy conservation in WSN through multilevel data reduction scheme, microprocessors and microsystems. http://dx.doi.org/10.1016/j.micpro.2015.05.019
6. Carvalho, C., Gomes, D. G., Agoulmine, N., & De Souza, J. N. (2011). Improving prediction accuracy for WSN data reduction by applying multivariate spatio-temporal correlation. *Sensors, 11*(11), 10010–10037.
7. Alduais, N. A. M., Abdullah, J., Jamil, A., & Heidari, H. (2017). Performance evaluation of real-time multivariate data reduction models for adaptive-threshold in wireless sensor networks. *IEEE sensors letters, 1*(6), 1–4.
8. Tayeh, G. B., Makhoul, A., Perera, C., & Demerjian, J. (2019). A spatial-temporal correlation approach for data reduction in cluster-based sensor networks. *IEEE Access, 7*, 50669–50680.
9. Dias, G. M., Bellalta, B., & Oechsner, S. (2017). The impact of dual prediction schemes on the reduction of the number of transmissions in sensor networks. *Computer Communications*. https://doi.org/10.1016/j.comcom.2017.08.002
10. Santini, S., & Romer, K. (2006). An adaptive strategy for quality-based data reduction in wireless sensor networks. In *Proceedings of the 3rd International Conference on Networked Sensing Systems (INSS 2006)*. Chicago, IL: TRF.
11. Liu, Y., Liu, C. X., & Zeng, Q. A. (2016). Improved trust management based on the strength of ties for secure data aggregation in wireless sensor networks. *Telecommunication Systems, 62*(2), 319–325.
12. Rajesh, G. (2016). Trust based temporal data aggregation for energy constrained wireless sensor networks. Doctoral dissertation, Anna University, Chennai.
13. Kumar, M., & Dutta, K. (2016). LDAT: LFTM based data aggregation and transmission protocol for wireless sensor networks. *Journal of Trust Management, 3*(1), 2.
14. Gilbert, E. P. K., Kaliaperumal, B., Rajsingh, E. B., & Lydia, M. (2018). Trust based data prediction, aggregation and reconstruction using compressed sensing for clustered wireless sensor networks. *Computers & Electrical Engineering, 72*, 894–909.
15. Gao, Y., Li, X., Li, J., & Gao, Y. (2018, October). A trustworthy data aggregation model based on context and data density correlation degree. In *Proceedings of the 21st ACM International Conference on Modeling, Analysis and Simulation of Wireless and Mobile Systems* (pp. 265–273). New York, NY: ACM.

16. Karthik, N., & Ananthanarayana, V. S. (2016). Data trustworthiness in wireless sensor networks. In *2016 IEEE Trustcom/BigDataSE/ISPA*. Piscataway, NJ: IEEE.
17. Karthik, N., & Ananthanarayana, V. S. (2017). Sensor data modeling for data trustworthiness. In *2017 IEEE Trustcom/BigDataSE/ICESS*. Piscataway, NJ: IEEE.
18. Karthik, N., & Ananthanarayana, V. S. (2017). Data trust model for event detection in wireless sensor networks using data correlation techniques. In *2017 Fourth International Conference on Signal Processing, Communication and Networking (ICSCN)*. Piscataway, NJ: IEEE.
19. Karthik, N., & Ananthanarayana, V. S. (2017). A hybrid trust management scheme for wireless sensor networks. *Wireless Personal Communications, 97*(4), 5137–5170.
20. Karthik, N., & Ananthanarayana, V. S. (2019). Trust based data gathering in wireless sensor network. *Wireless Personal Communications*: 1–21.
21. Kumar, R. K., Kumar, G. P., & Sakthivelmurugan, V. (2017). Predicting and recovering link failure localization (PRLFL) in wireless sensor networks. *Asian Journal of Research in Social Sciences and Humanities, 7*(2), 851–864.
22. Dias, G. M., Bellalta, B., & Oechsner, S. (2016). A survey about prediction-based data reduction in wireless sensor networks. *ACM Computing Surveys, 49*(3), 58.
23. Samuel, M. (2004). Intel lab data 2004–2006. http://db.csail.mit.edu/labdata/labdata.html

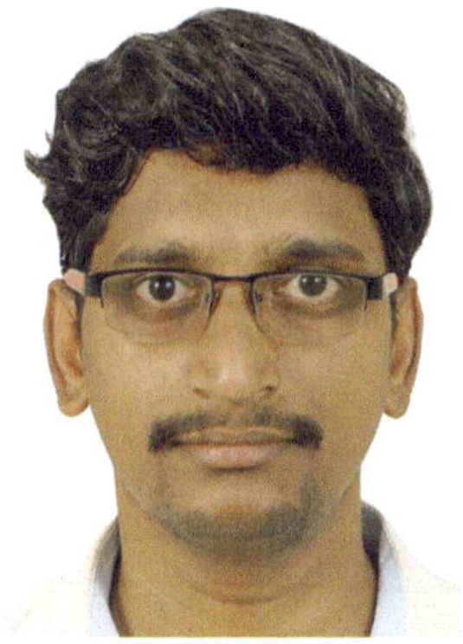

N. Karthik is currently working as Assistant Professor in the Department of Computer Science and Engineering, Karunya Institute of Technology and Sciences, Coimbatore, Tamilnadu, India. He is pursuing PhD at Department of Information Technology, National Institute of Technology Karnataka, India. He completed his graduation in Computer Science and Engineering, from Anna University, Chennai and post-graduation in Pervasive Computing Technology, from Anna University of Technology, Tiruchirappalli. His research interest includes Pervasive Computing and Semantic Web.

V. Ebenezer received his B. Tech degree in Information Technology and M.E degree in Computer Science and Engineering from Anna University, Chennai in the year 2009 and 2012, respectively. He is pursuing PhD at Anna University, Chennai. He is currently working as Assistant Professor in the Department of Computer Science and Engineering, Karunya Institute of Technology and Sciences, Coimbatore, Tamilnadu, India. He has published research papers in the various International Conferences and Journals. His research interest includes Cloud Computing, Body Area Networks, Data Structures and Distributed Systems.

Prospects of Sustainable Transport of Automobile Vehicles in Kolkata City

Sushobhan Majumdar

1 Introduction

Recently, environmental conditions and problems is one of the major parts of environmental concerns regarding the sustainable development of an area. Sustainable transport is one of the important aspects in the sustainable development studies. In case of developing countries, the number of automobile vehicles has been increasing day by day, which creates adverse impacts on the local land resources. In case of developing countries like India, the major problems are the vehicles which are very old in nature and their level of pollution is also high. Chan et al. [1] describes that it is source of environmental problems and health problems as in most cases it is related with the noise pollution, air pollution, etc. According to the Frank et al. [2], if the number of automobile vehicles is increased then it will impact physical and mental health of the human beings and it is the sources of various chronic diseases. To reduce the problems regarding the problems of transport, various new technologies should be introduced for the future planning and development of the region. Especially in the developed countries, the demand for the motorized vehicles is higher than the developing countries because of increasing population pressure of the area. To cope up with the huge growth population with the increase in the motorized vehicles different environmental scientists have planned various policies regarding the sustainable transportation and environmental planning to improve the quality of life of the city people [3, 4]. Kolkata city is one of the cities in the eastern part of India, which has been experienced with the various modes of transport and it is also the regional headquarters of various developmental strategies.

S. Majumdar (✉)
Department of Geography, Jadavpur University, Kolkata, India

G. R. Kanagachidambaresan et al. (eds.), *Internet of Things in Smart Technologies for Sustainable Urban Development*, EAI/Springer Innovations in Communication and Computing, https://doi.org/10.1007/978-3-030-34328-6_5

Miller and Shaw [5] and Miller [6] have used some analytical issues regarding the uses of Geographical Information System (GIS) in the transportation sector. GIS is basically collecting, managing, and analyzing the spatial data in an easy way [7, 8]. GIS helps in integration of spatial data with the coding and management of the data [9]. Most of the previous researchers have used GIS for exploring information and analyzing like air pollution analysis, noise pollution etc. [10–12]. In Kolkata city, the demand for the public transport is very much higher than the other metro cities in India. The objective of this study is that to find out the prospects of sustainable transport in Kolkata city and suggest few strategies to make Kolkata city more sustainable in case of transportation sector.

2 Study Area

Kolkata is one of the major cities in the eastern part of India which have experienced with one of the largest urban agglomerations in the country. Kolkata city is situated over the large Gangetic delta of the river Hooghly [13]. Kolkata was the capital of British India till 1916. Kolkata city has been flourished from 1970s which is mainly because of the influx of refugees (from East Pakistan or Bangladesh), good infrastructural facilities, good accessibility from the other areas etc. The central part of Kolkata city is regulated by the Kolkata Municipal Corporation. The surrounding areas or metropolis of Kolkata city is known as the Kolkata Metropolitan Area (KMA). Kolkata Metropolitan Area (Fig. 1) is under the jurisdiction of Kolkata Metropolitan Development Authority (KMDA). All types of planning and development in this area are under the rules and regulations of KMDA. The population of Kolkata Metropolitan Area is 15.89 million (according to the latest census) with a growth rate of 10.30 per cent. KMA is consisting of different units like municipal corporations, municipalities, census towns, out growth and rural mouzas or villages. There are three municipal corporations and 42 municipalities in the KMA. The rate of population growth is high in the fringe area than the CBD area of the Kolkata city. Kolkata city is consisting of various modes of transport like railway (broad gauge), metro railway, bus transport, auto rickshaw transport, taxi, tram service, ferry service, etc. These transport modes are the lifeline of Kolkata city as most of the people from the fringe areas or peripheral areas of the city used to go to Kolkata city for their job purposes. For this reason behind the growth and development of Kolkata city, transport sector plays a vital role (Table 1).

3 Data and Methods

Transport and its associated problems is a serious issue in the every cities of the world especially where the population pressure is high. Kolkata is one of the cities among them which have facing various problems related to transport. To find out the

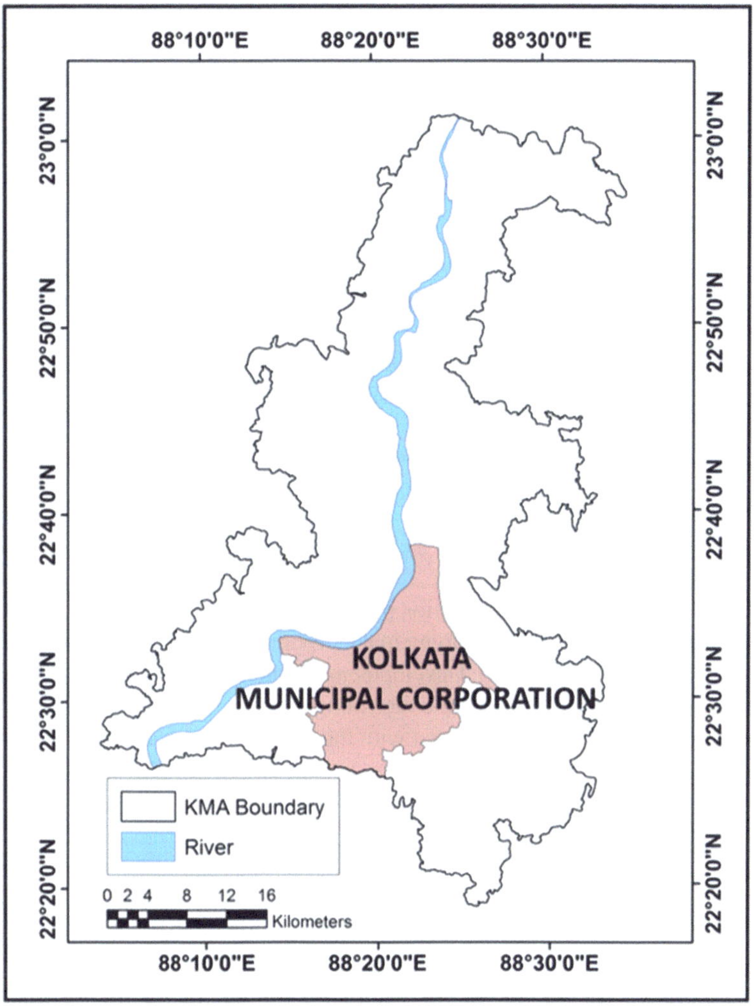

Fig. 1 Areas of study

various problems of Kolkata city, different remote sensing and GIS techniques have been used. For the analysis of surface temperature and vegetation cover, landsat satellite images of 2019 have been used. To find out the land surface temperature and vegetation of Kolkata city, following parameters have been used.

$$L\lambda = \text{MLQcal} + \text{AL} - O\text{i (For Landsat OLI 8)}$$

where Lλ = TOA spectral radiance (Watts/(m^2 × srad × μm)), ML = band-specific multiplicative rescaling factor from the metadata (RADIANCE_MULT_BAND_x, where x is the band number), AL = band-specific additive rescaling factor

Table 1 Transport in Kolkata metropolitan area

Modes of transport	Name of transport
Road network	Bus, taxi, auto rickshaw, two wheeler, toto (especially peripheral areas, cab service)
Rail network	Eastern railway (broad gauge), metro railway, circular railway
Ferry service	Mainly in the Hooghly river

from the metadata (RADIANCE_ADD_BAND_x, where x is the band number), QCal = quantized and calibrated standard product pixel values (DN), and *O*i is the correction for band 10.

Normalized Difference Vegetation Index (NDVI) is mainly used to find out the vegetation coverage of an area. It is generally expressed as

$$\text{NDVI} = \frac{\text{Band 5} - \text{Band 4}}{\text{Band 5} + \text{Band 4}} \text{ (For Landsat OLI 8)}$$

NDVI is related to variables such as leaf area, canopy coverage, and chlorophyll density [14].

The data regarding the transportation problems of Kolkata city has been collected from the Kolkata police, Kolkata Municipal Corporation Office, Transportation data from the Kolkata Metropolitan Development Authority, etc. To analyze various transportation problems like traffic congestion, road accidents, etc., different cartographical techniques have been used with the help of statistical analysis.

4 Results and Analysis

Kolkata city is one of the major cities in the eastern part of India. Kolkata is one of the cities in India which is consisting of old hand pulled rickshaw to modern metro railway transport. Though the sphere of tram service has been reduced because of the traffic congestion in some areas, but it is still maintained his service (Fig. 2).

Kolkata city is one of the cities consisting of different types of road networks like national highway, state highway, major road, minor road, metalled road, unmetalled road, etc. There are few national highway and state highway in the Kolkata city. The number of metalled road is high towards the Kolkata city which decreases outwards and the number of un-metalled roads is high in the peripheral areas of the Kolkata city which decreases towards the city areas. After the visual observation of satellite images around Kolkata city and from the field verification, it has been found that the growth and expansion of Kolkata city is mainly transit or transport oriented development. In most cases, it has been found that most of the settlement towards the peripheral areas of Kolkata city has grown through a linear way along the major transportation routes. These tendencies have given birth to the urban sprawl which indirectly accelerates negative externalities of Kolkata city.

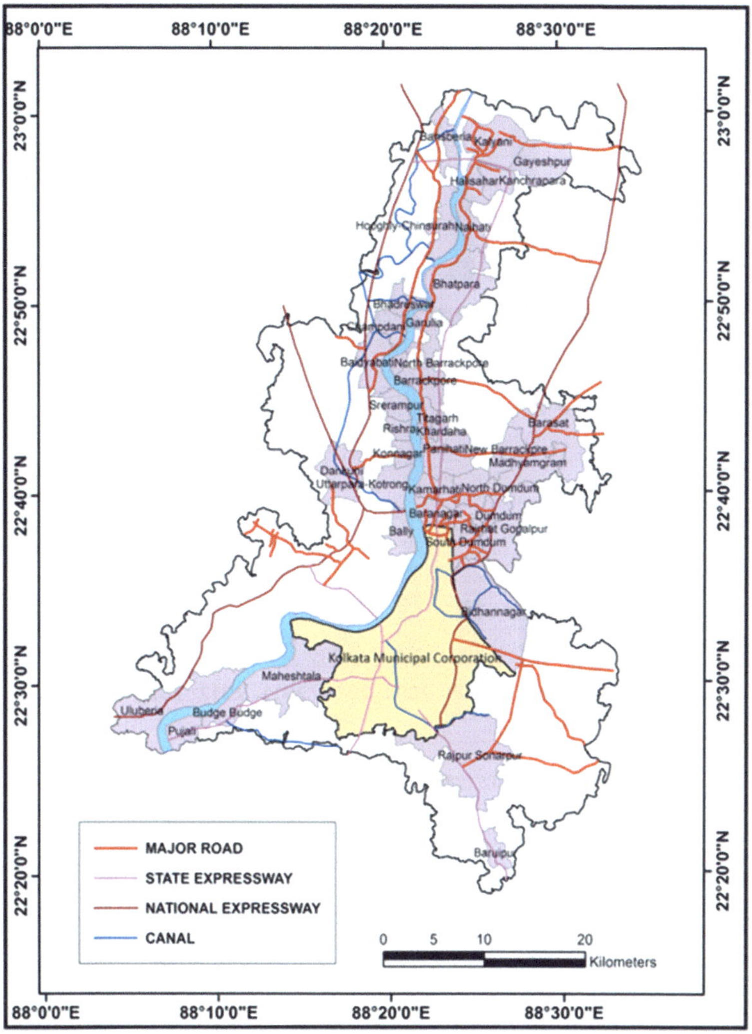

Fig. 2 Transport network in Kolkata metropolitan area

From various governmental reports regarding the number of automobile vehicles, it has been found that the number of automobile vehicles in Kolkata city is growing very fast. The numbers of automobile vehicles are maximum after 2012 (Fig. 3), which creates extreme pressure on the transport condition of the city.

From the data regarding the number of registered vehicles (cars) of Kolkata city, it has been found that the number of cars in Kolkata city is highly increased and it tends very high after 2012, which creates pollution in Kolkata city (Figs. 3 and 4).

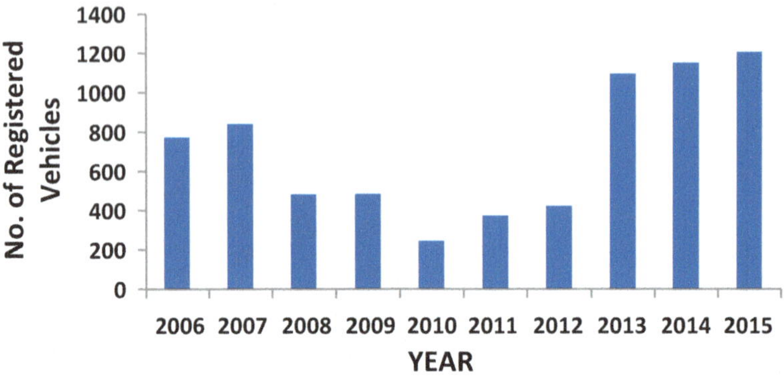

Fig. 3 Number of registered vehicles in Kolkata city

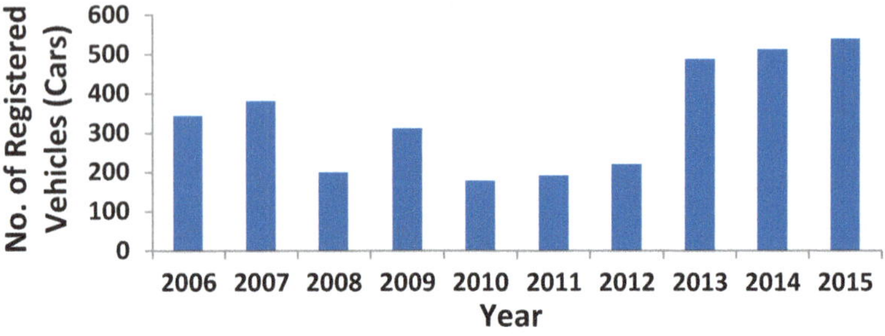

Fig. 4 Number of registered vehicles (cars) in Kolkata city

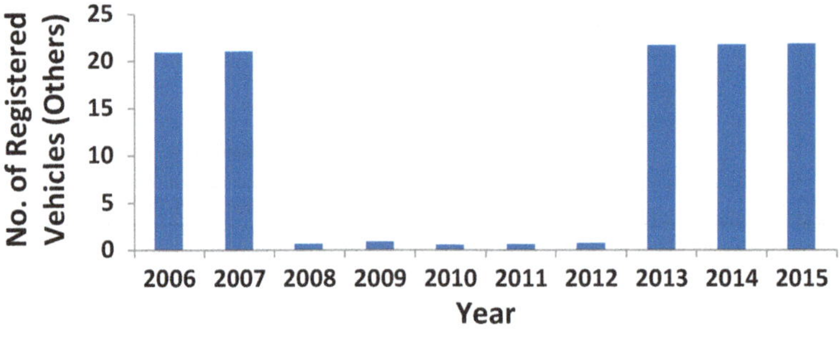

Fig. 5 Number of registered vehicles (others) in Kolkata city

Figure 5 describes about vehicles mainly used for other purposes in Kolkata city. In this case, the vehicles used for other purposes have been used. The number of registered vehicles mainly used for other purposes has been hugely increased in the last 5 years.

The number of two wheelers has been increased during the last few years, though the number of registered vehicles is almost same throughout the period. But from the

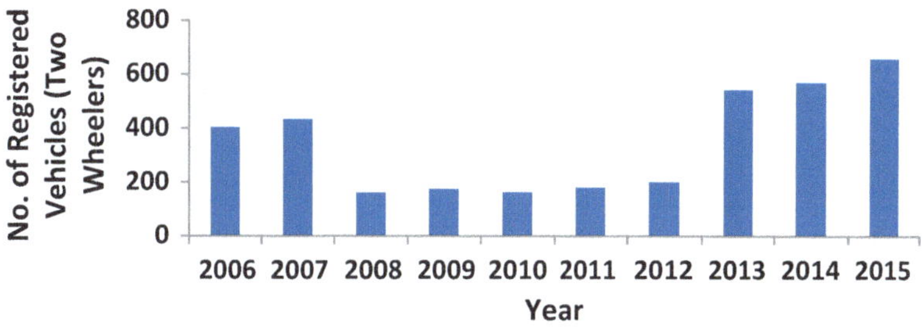

Fig. 6 Number of registered vehicles (two wheelers) in Kolkata city

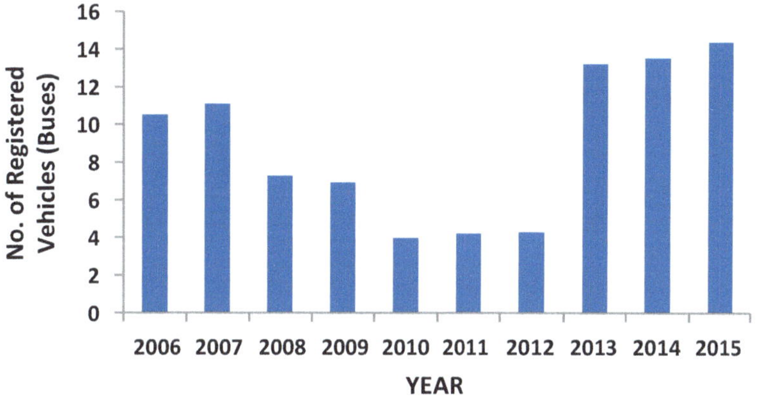

Fig. 7 Number of registered vehicles (buses) in Kolkata city

year 2013 it increases and is maximum in 2015. In the fringe and peripheral areas, the development is mostly unplanned. So, huge growth of automobile vehicles is creating various problems like traffic congestion, road accidents, etc. (Fig. 6). For this reason negative externalities of transport in Kolkata city has been increasing day by day.

Because of the increase in the number of passengers, the services of the buses of various transport agencies like Calcutta State Transport Corporation (CSTC) and South Bengal State Transport Corporation (SBSTC) has been increased. Increase in the number and service of the public bus reduces congestion and traffic problems in Kolkata city during peak hours. So, the number of buses has been increased in recent decades (Fig. 7).

Figure 8 describes the number of registered light vehicles in Kolkata city which increases in recent time periods. Figure 9 discusses about the growth of taxis in Kolkata city which has been increased from the past few years which because of the various Governmental policies like "Gatidhara" and introduction of various cab services like (Ola, Uber, etc.)

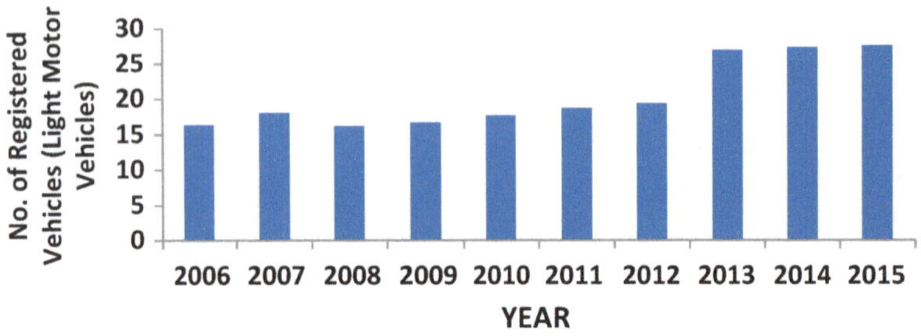

Fig. 8 Number of registered vehicles (light motor vehicles) in Kolkata city

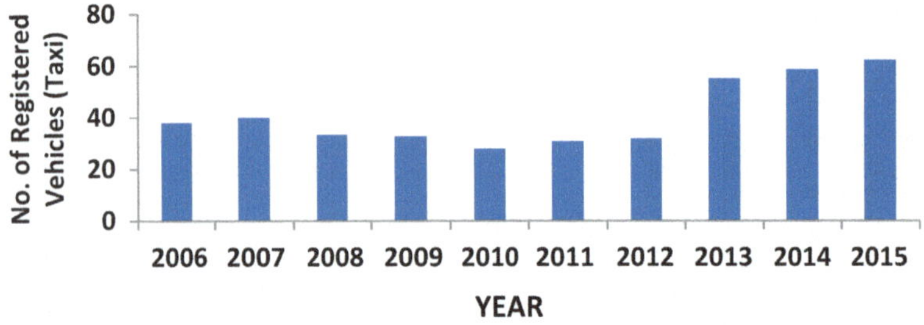

Fig. 9 Number of registered vehicles (taxi) in Kolkata city

Because of the huge increase in the automobile vehicles of Kolkata city, the temperature of Kolkata city has been increased in the recent years.

From the land surface temperature of Kolkata city, it has been found that the air temperature or surface temperature of Kolkata city is very high in the core areas of Kolkata city because of availability of lots of metalled roads and huge number of automobile vehicles (Fig. 10). The temperature is very high especially in the Dalhousie area, Dharmatala area, and Parkstreet area. Land surface temperature decreases towards the outer areas from Kolkata city as the density of roads and number of vehicles decreases and also the pressure of population decreases. Because of the huge temperature increase in the CBD areas of Kolkata city, this city has now transformed into a heat island situation which impacts on the urban sustainability of Kolkata city.

Figure 11 Shows the condition of vegetation in Kolkata city and it has also been found that in the central part of Kolkata city the temperature is very high and the number of vegetation is very low which negatively impacts on the urban ecology of the city. Because of the low amount of vegetation, high density of roads, and maximum number of automobile vehicles, the temperature is high in the city core areas, which creates adverse impacts on the local land resources.

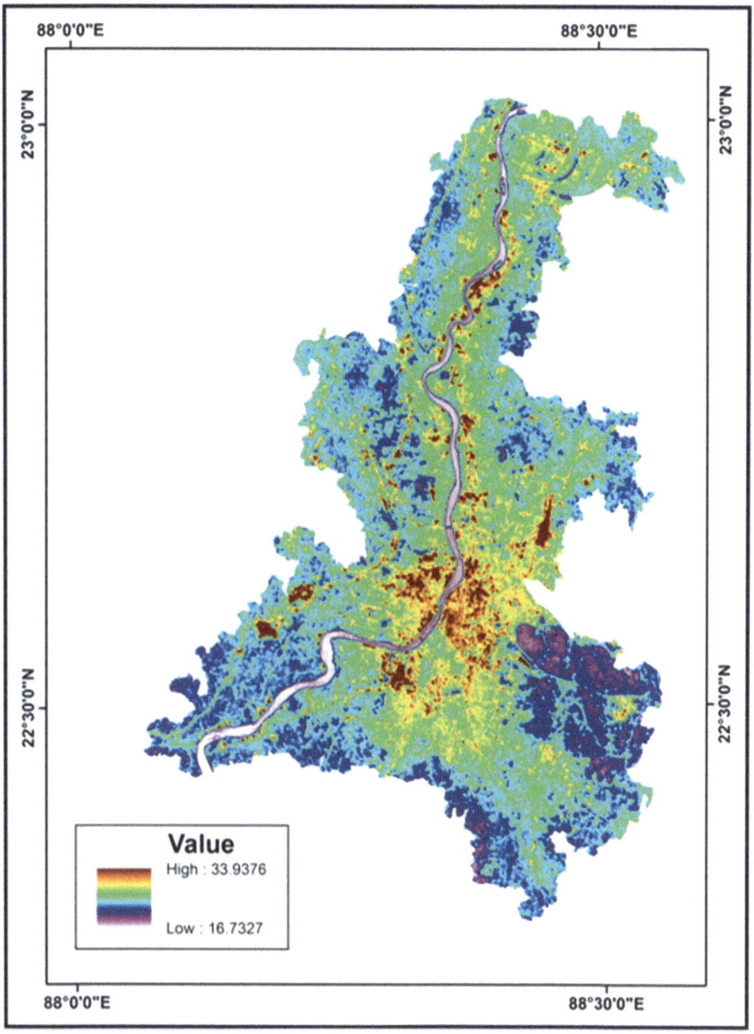

Fig. 10 Land surface temperature of Kolkata city

5 Discussions

From the various reports regarding the automobile vehicles in Kolkata city, it has been found that number of automobile vehicles in Kolkata city has been increased in recent decades, which create extreme pressure on local land resources. The width of the road is very slightly increased but the number of vehicles has been hugely increased, which creates extreme pressure on transport condition of the area which accelerates negative externalities of transport. Another noticeable fact is that to

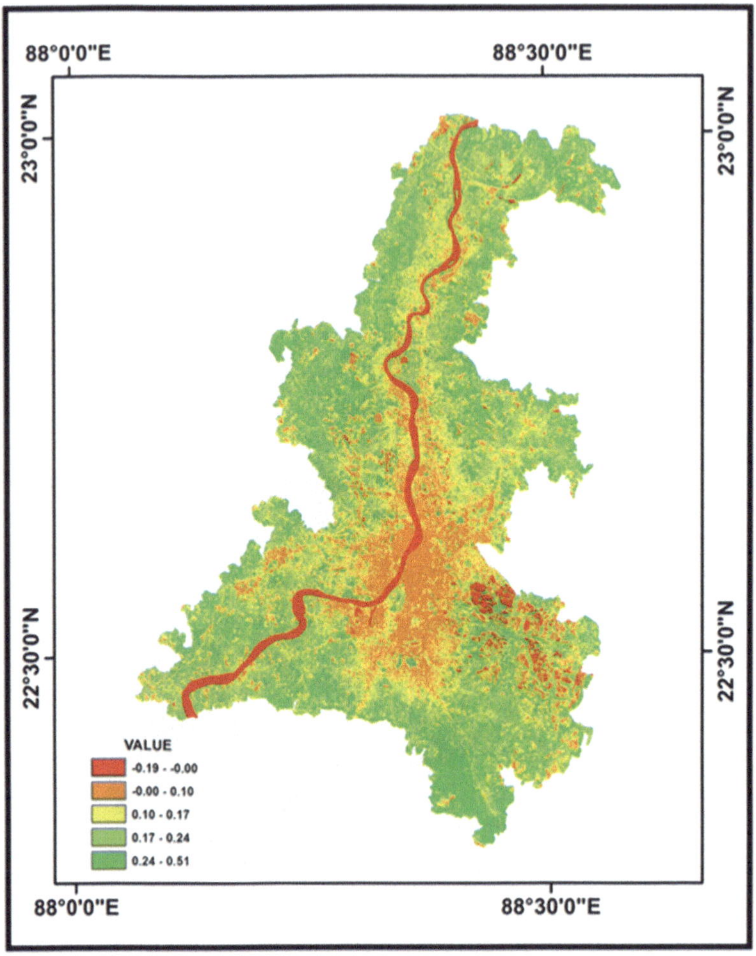

Fig. 11 Normalized difference vegetation index of Kolkata city

reduce the passenger pressure of auto rickshaw services, toto services have been started which creates huge traffic congestion in the peripheral areas. Another fact regarding toto services in Kolkata city is that there are no proper registration systems to control the services of the toto. For this reason the problem of traffic congestion has been flourished also in the peripheral areas of Kolkata city. Though the Govt. officials and policy makers have taken initiative to reduce the problem of toto services, it is insufficient to eradicate all the problems.

6 Problems Related to Automobile Vehicles in Kolkata City

From the detailed study over the transportation scenario of Kolkata city, it has been found that Kolkata city is now facing lots of problems. Those are as following.

The main problem of Kolkata city is the huge population growth which is mainly influx of population both from the surrounding regions and also from the other countries. As a result, population pressure in each area has been increased. Road and passenger ratio has been increased because the rate of demographic growth is very high than the widening and development of roads. In the CBD areas, the width of the roads is very low, which creates problems to the bus drivers or heavy goods carriers to move. So, during peak hours it creates huge traffic congestion. Except CBD areas and few highways the condition of road is very bad which creates traffic accidents regularly. Auto rickshaw transport is another life line of Kolkata city, but it is also source for the huge traffic congestion during peak hours. Un-fixed fare and bad behavior of auto rickshaw drivers are also a problem. Another problem regarding the auto rickshaw transport is though auto rickshaw in the city areas is Liquid Petroleum Gas (LPG) operated, but in the south western fringe areas (especially in the Pujali areas) it is diesel operated which is not eco-friendly and causes a lots of air pollution. These are the major problems of transport in Kolkata city.

7 Recommendations

In case of Kolkata city, the problem of transportation is so massive, and it is the areas of concern of urban planners and decision makers. In case of the making of policy, the causes and effects of the problem must be considered. In few cases, the policy makers may consult with state governments, national governments, and the local people. The following strategies can be taken.

7.1 Pricing and Taxation

Pricing and taxation are two important issues related with transport. Though both the terms are different, they are related with each other. Government officials and policy makers can increase the price of the cars (four wheelers) or bikes. On the other hand, they can decrease the prices of the public vehicles by giving subsidies. Because of high pricing, the person belonging from the middle income groups became unable to purchase it. So they have to travel by public transport. Similarly by increasing taxation, the government officials can regulate the growth of public vehicles easily. Less number of cars in the road will minimize the level of pollution in Kolkata city, which will affect on the urban sustainability.

7.2 Planning

By the proper land use and transportation planning, the authority can easily minimize the level of pollution in Kolkata city. Land use and transportation planning can be interlinked with each other. The traffic congestion can be decreased by advanced automatic traffic controlling system. The property tax or tax related to land can be lowered in the areas around the major roads. As a result, people can easily travel with the public transport. Special trafficking system can be introduced during peak hours. In the congested areas during the peak hours entering the huge goods carrier should be strictly banned.

7.3 Policy

If the speed of the automobile vehicle increases, then it will consume more fuel and vice versa. So, new technologies must be introduced which will consume less fuel. Though government of India has introduced i3s technologies for less fuel consumption in bike, it is insufficient to solve this problem. Except these, policy makers should focus on the alternative modes of energy except fuel because of its limited stock. Other than these, government officials should focus on the battery-operated auto rickshaw which is eco-friendly in nature. In some cases, following measures can also be taken.

- Regular checkup of pollution level.
- Increasing tax in private transport and fuel-operated transport.
- Subsidies for new technologies and alternative modes.
- Wider roads and advanced traffic system.

Except these different sustainable demand management policies may be taken like vehicle retirement, car pooling, car sharing, parking restraints, eliminating company cars, etc.

7.4 Education

Different transport-related studies may be introduced such as the environmental impacts of automobiles transport, transportation planning, transportation management, driver education, the merits of bicycle riding, the merits of transit and subsidies, etc. This type of study will increase the awareness among people to eradicate the problems of automobile vehicles in Kolkata city.

7.5 Technology

The government of India also introduced some new less fuel consuming vehicles. New technologies (like i3s) can be adopted. Some of the new technologies Advanced Traffic Management System (ATMS), Advanced Vehicle Control System (AVCS), Advanced Public Management System (APMS), Commercial Vehicle Operations (CVO), and Advanced Rural Transportation System (ARTS) can be adopted.

8 Conclusion

From the above analysis, it has been found that the problems of Kolkata city due to excessive growth of automobile vehicles are increasing day by day. Excessive growth of automobile vehicles in the city core areas creates different problems like congestion, temperature increase, traffic accidents, air pollution, etc. Because of the huge traffic congestion, the land surface temperature in the city core areas is increasing day by day, which creates Kolkata city a heat island like situation. The temperatures in the city core areas are higher than the peripheral areas of the city. This creates extreme adverse impact on local ecosystem. Due to the traffic congestion, the number of accidents has also been increased, which impacts on the urban sustainability of Kolkata city. To minimize the problem and regulate the growth, different strategies may be taken, which will regulate the growth of automobile vehicles. Different new technologies like i3s technologies can be introduced. Government should encourage people to use public transport rather than the private transport. Government officials should encourage battery- or electric-operated automobile vehicles rather than operated by fossil fuels. Other than these alternative, fuels may be introduced. This study is helpful not only to the transportation researchers, but also it is helpful to the other policy makers also. It will throw a new light on the future transportation researchers in the field transport oriented research. The method and techniques used in this study in case of Kolkata city can also be applied in the major or minor cities also.

References

1. Chan, L., Lau, W., Zou, S., Cao, Z., & Lai, S. (2002). Exposure level of carbon monoxide and respirable suspended particulate in public transportation modes while commuting in urban area of Guangzhou, China. *Atmospheric Environment, 36*, 5831–5840.
2. Frank, L. D., Sallis, J. F., Conway, T. L., Chapman, J. E., Saelens, B. E., & Bachman, W. (2006). Many pathways from land use to health: Associations between neighborhood walkability and active transportation, body mass index, and air quality. *Journal of the American Planning Association, 72*, 75–87.

3. Ma, L., & Dill, J. (2015). Associations between the objective and perceived built environment and bicycling for transportation. *Journal of Transport and Health, 2*, 248–255.
4. Ülengin, F., Kabak, Ö., Önsel, Ş., Ülengin, B., & Aktaş, E. (2010). A problem-structuring model for analyzing transportation–environment relationships. *European Journal of Operational Research, 200*, 844–859.
5. Miller, H., & Shaw, S.-L. (2001). *Geographic information systems for transportation: Principles and applications*. New York: Oxford University Press.
6. Miller, H. J. (1999). Potential contributions of spatial analysis to geographic information systems for transportation (GIS-T). *Geographical Analysis, 31*, 373–399.
7. ESRI. (2017). *What is GIS?* Retrieved April 25, 2019, from http://www.esri.com/what-is-gis
8. USEPA. (2011). *Guide to sustainable transportation performance measures* (pp. 1–55). Washington, DC: USEPA.
9. Al-Ali, M. M. (2016). Applicability of GIS tools in assessing performance of the transportation systems in urban areas. *World Journal of Science, Technology and Sustainable Development, 13*, 120–127.
10. Armstrong, J. M., & Khan, A. M. (2004). Modelling urban transportation emissions: Role of GIS. *Computers, Environment and Urban Systems, 28*, 421–433.
11. Zytoon, M. A. (2016). Opportunities for environmental noise mapping in Saudi Arabia: A case of traffic noise annoyance in an urban area in Jeddah City. *International Journal of Environmental Research and Public Health, 13*(5), 496.
12. Barsi, J. A., Schott, J. R., Hook, S. J., Raqueno, N. G., Markham, B. L., & Radocinski, R. G. (2014). Landsat-8 thermal infrared sensor (TIRS) vicarious radiometric calibration. *Remote Sensing, 6*(11), 11607–11626.
13. Nair, T. (1986). *Calcutta in the 17th century*. Calcutta: Firma KLM.
14. Weiss, J. L., et al. (2004). Long-term vegetation monitoring with NDVI in a diverse semi-arid setting, central New Mexico, USA. *Journal of Arid Environments, 58*, 249–272.

Sushobhan Majumdar is an Applied Geographer with Post Graduate Degree from Presidency University, Kolkata and Doctoral Degree from the Department of Geography, Faculty of Science, Jadavpur University, Kolkata. He is currently working on research on the Geography of Transport with the regional planning and development. His principal area of research includes Urban Geography, Social Geography, Regional Development and Planning, Remote Sensing, and GIS. He is also engaged in the various developmental projects under the Socio Economic Planning Unit of Kolkata Metropolitan Development Authority (KMDA) under the Government of West Bengal. He has published 36 research papers in different journals, proceedings, and edited volumes of national and international repute. He has presented papers in more than 20 national and international seminar and conferences in India and abroad. He delivered special or invited lectures in various national and regional programs. He is member of Geographical Society of India, National Association of Geographers of India (NAGI), Institute of Indian Geographers (IIG), Regional Science Association of India (RSAI), etc. Recently, he is deeply engaged in surface temperature mapping and modeling of metro cities in India and its impact on environment.

The Application Scenario and Dependence with IoT

Kavita Suryawanshi and Manasi Pandharkar

1 Introduction

The chapter has an academic contribution considering the emerging trends in the field of computer and information technology, especially IoT a very vibrant research topic. India is at the forefront of this digitalization with government initiatives such as Digital India, and IoT is the key enabler for this digital transformation.

Internet of Things (IOT) is networking of physical devices (also referred as connected devices and smart devices) embedded with electronics, software, sensors, actuators, and network connectivity that enable these objects to collect and exchange data [1].

The International Telecommunication Unit defined IOT as "a global infrastructure for the information society enabling advanced services by interconnecting things based on existing and evolving interoperable information and communication technologies" [2].

IOT is network of internet connected objects (things) able to collect and exchange data. Internet of Things will be as transformative to the world as was an Industrial Revolution. IoT is classified as consumer IoT and industrial IoT. There are many technologies that enabled IoT to become popular like sensors' prices have dropped, cost of bandwidth has reduced, processing cost has declined, and IPv6 is enabled to connect more devices to the internet, communication protocols, and cloud technology [3].

K. Suryawanshi (✉) · M. Pandharkar
D. Y. Patil Institute of MCA and Management, Pune, India

Savitribai Phule Pune University, Pune, India

G. R. Kanagachidambaresan et al. (eds.), *Internet of Things in Smart Technologies for Sustainable Urban Development*, EAI/Springer Innovations in Communication and Computing, https://doi.org/10.1007/978-3-030-34328-6_6

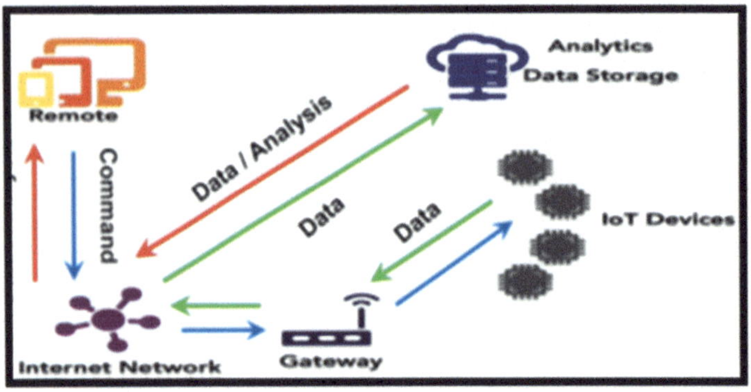

Fig. 1 Working of Internet of Things

1.1 Working of IoT

Devices and objects with built-in sensors are connected to an Internet of Things platform, which integrates data from the different devices and applies analytics to share the most valuable information with applications built to address specific needs. These powerful IoT platforms can pinpoint exactly what information is useful and what can safely be ignored. This information can be used to detect patterns, make recommendations, and detect possible problems before they occur. For example, if a person owns a car manufacturing business, he might want to know which optional components (leather seats or alloy wheels, for example) are the most popular. Using Internet of Things technology, he can

- Use sensors to detect which areas in a showroom are the most popular, and where customers linger longest.
- Drill down into the available sales data to identify which components are selling fastest.
- Automatically align sales data with supply, so that popular items do not go out of stock.

The information picked up by connected devices enables him to make smart decisions about which components to stock up on, based on real-time information, which helps him save time and money [3].

Figure 1 depicts the working of Internet of Things. There are different IoT devices such as sensors, actuators through which data is sensed and via gateway it goes on Internet. These sensed data is analyzed and get stored on cloud storage. Using remote communication channels, the analyzed data will get distributed to the destinations.

In a nutshell, the Internet of Things is the concept of connecting any device (so long as it has an on/off switch) to the Internet and to other connected devices. The

IoT is a giant network of connected things and people—all of which collect and share data about the way they are used and about the environment around them [4].

That includes an extraordinary number of objects of all shapes and sizes—from smart microwaves, which automatically cook your food for the right length of time, to self-driving cars, whose complex sensors detect objects in their path, to wearable fitness devices that measure your heart rate and then number of steps you've taken that day, then use that information to suggest exercise plans tailored to you. There are even connected footballs that can track how far and fast they are thrown and record those statistics via an app for future training purposes [5].

2 IoT in Industry Applications

The use of the IoT is spread across all the verticals of the industry. From logistic to supply chain, from controlling to security everywhere IoT is used. For making things easier, efficient, faster, and smart, IoT is the best choice for the industry. IIOT is used in the automation accompanies to improve their traditional business processes to make it smarter. IIoT is the technological era which is changing the face of the industry automation. What exactly the IIoT is?

Industry experts and market analysts define the IIoT as "The Industrial Internet of Things (IIoT) is the next wave of innovation impacting the way the world connects and optimizes machines. The IIoT, through the use of sensors, advanced analytics and intelligent decision making, will profoundly transform the way field assets connect and communicate with the enterprise" [6].

"Leading O&G companies are building an infrastructure where sensors, data management, advanced analytics and automation are being used to unlock production, reduce operating costs and optimize assets."

"The Industrial Internet, a connected network of intelligent machines working the way they are intended, will transform business as dramatically as the consumer Internet has changed our lives."

In Fig. 2, at the bottom of the stack are the devices (systems) on the factory or process floor. These can be field sensors, controllers, or PCs, and all of these hardware systems can include (or not) aspects of hardware security. These end devices must have useful data to communicate and are generally connected to communication hubs, gateways, and switches, so that the data is put in the cloud (or an intranet) as big data.

Once this data is out "there," different analytics and optimization software can be developed to optimize the manufacturing assets for a myriad assortment of tasks: system uptime, scheduled maintenance, power efficiency, and more efficient resource utilization.

But that is not all. The IIoT promises that this data can be integrated within the manufacturer's ERP and CRM software. The manufacturing operation can not only be used to plan and cost out manufacturing processes more efficiently, but even to

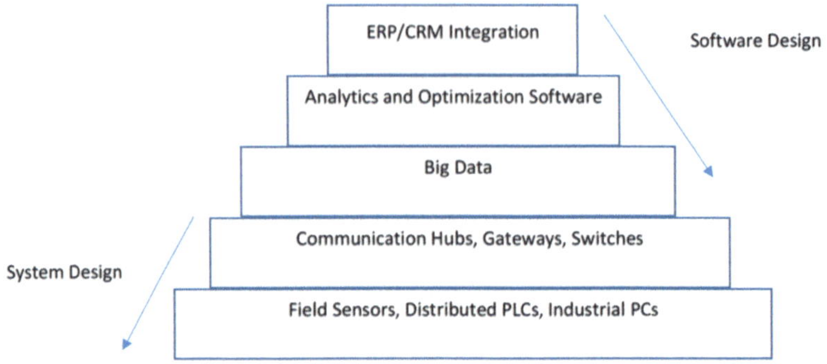

Fig. 2 Industrial Internet of Things [6]

use customer information to change assembly lines and process parameters in real time.

The bottom two stacks in Fig. 1 impact the design of system hardware, and the top layers affect software development and integration.

While the long-term impact of IIoT is at times difficult to predict, three distinct operational environments will set the stage for the smart manufacturing enterprise to emerge.

- **Smart Enterprise Control**—IIoT technologies will enable tight integration of smart connected machines and smart connected manufacturing assets with the wider enterprise. This will facilitate more flexibility and efficiency, and hence profitable, manufacturing. Smart enterprise control can be viewed as a mid-to-long-term trend. It is complex to implement and will require the creation of new standards to enable the convergence of IT and OT systems.
- **Asset Performance Management**—Deployment of cost-effective wireless sensors, easy cloud connectivity (including WAN), and data analytics will improve asset performance. These tools allow data to be easily gathered from the field and converted into actionable information in real time. This will result in better business decisions and forward-looking decision-making processes.
- **Augmented Operators**—Future employees will use mobile devices, data analytics, augmented reality, and transparent connectivity to increase productivity. As fewer skilled workers are left behind to man core operations due to a rapid increase in babyboomer retirement, younger replacement plant workers will need information at their fingertips. That information will be delivered in a real-time format that is familiar to them. Thus the plant evolves to be more user-centric and less machine-centric [6].

2.1 *IIoT in Manufacturing Industry*

In 2016, manufacturing operations alone accounted for an IoT spend of $102.5 billion on a total of $178 billion, and all IoT use cases in manufacturing are combined. With a total spend of $178, manufacturing overall is by far the largest industry in the Internet of Things AND of the Industrial IoT and the segment of manufacturing operations outweighs all other IoT use case investments across all industries, consumer included [7].

Two other IoT use cases which are important in manufacturing from a spending perspective, on top of operations, are production asset management and maintenance and field service, according to the mentioned research by IDC, released early 2017 [8].

In the age of Industry 4.0 and the digital transformation of manufacturing, the manufacturing industry is the market where most Industrial IoT (IIoT) projects are realized and by far the market where most IIoT investments are made.

IoT is a core component of industrial transformation efforts across the globe, including Industry 4.0 (with its fourth industrial revolution) and the Industrial Internet (with the Industrial Internet Consortium).

Moreover, manufacturing is not just the clear leader in the Industrial Internet but it tops ALL industries (including the consumer IoT space) in the broader IoT reality. According to IDC data, published early 2017, the manufacturing industry was good for a total IoT spend of $178 billion in 2016, which is more than twice as much than the second largest vertical market (in IoT spend), transportation.

2.1.1 IoT Use Cases in Manufacturing: Opportunities and Context

If we look at IoT in manufacturing from an IoT use case, we see that one use case is sticking out: manufacturing operations, which is also the largest use case across all industries globally.

Manufacturing obviously covers many types of products, operations, processes and a vast space of activities, components, machines, people, partners, information systems and so forth. It takes a long time to convert raw materials to finished goods and it includes supply chains, logistics, and transportation as well.

If we look at manufacturing as industrial production in any of the stages where raw materials are turned into products or products are used to build other products, it is clear that we see a huge market that is highly interconnected. It is among others in this sense that the Internet of Things almost by definition is key for the manufacturing industry in an integrated approach, and it includes the technologies such as big data analytics, cloud, and robotics and, most importantly perhaps, the integration of IT (Information Technology) and OT (Operational Technology).

Part of them are to be seen in the context of the digital/connected factory, others refer more to facility and asset management and still others relate to components such as safety/security and operations/logistics/ecosystems. Last but not least there is the customer preference and behavior part. Below is a list with several IoT use cases in manufacturing and their benefits/realities [8].

2.1.2 Manufacturing Operations

Manufacturing operations is the largest manufacturing IoT use case. It includes operations of manufacturing facilities, involving several assets and operational personnel.

Manufacturing operations include the several elements which are typical in Manufacturing Operations Management (MOM), such as asset management, intelligent manufacturing, performance optimization and monitoring, planning, human–machine interaction, end-to-end operational visibility, and these cyber-physical systems as we know them from Industry 4.0.

2.1.3 Production Asset Management and Maintenance

This is the second largest IoT use case in manufacturing and consists of a range of applications.

It includes production asset monitoring and tracking, from location to the monitoring of parameters in several areas such as quality, performance, potential damage or breakdowns, bottlenecks, the list goes on. On top of performance and optimization, there is of course also the dimension of maintenance (as a result and/or in a predictive way).

It is clear that asset management and maintenance in a manufacturing industry setting goes beyond pure production assets.

2.1.4 Field Service

According to the mentioned IDC report, this is the third most important IoT use case in manufacturing.

In this the manufacturer acts as a service provider. From product-related services to business-related services: the (field) service organizations of manufacturers are important drivers of growth. In the field service of manufacturing, planning, scheduling, and transportation of the products, IoT ecosystem plays the vital role [8].

2.1.5 Other Manufacturing IoT Use Cases

Other than all the three fields mentioned above which uses IoT, still there are some more areas which uses the IoT and IIoT concepts.

Safety, security, worker protection (and productivity), and the many links between manufacturing in the strictest sense with connected services/operations/ industries such as transport, supplier management, and so forth all contribute to the vast IoT-related manufacturing solutions.

Vehicle and asset tracking, connected factory applications, staff safety applications, health monitoring (real-time), smart ventilation and air quality management, smart environmental measurement, access control (security), smart measurement of presence/levels of liquids, gases, radiation and dangerous materials (depending on the type of operation), asset protection, facility management, risk measurement, etc. use the concept of IoT.

2.1.6 Visualizing the Internet of Things in Manufacturing

To visualize the usage of the Internet of Things in manufacturing with an overview of the various mentioned places/contexts where IoT is leveraged in the manufacturing space, Fig. 3 below from Microsoft offers a great overview.

It distinguishes between the manufacturing plant, global facility insight (beyond the factory), the customer site, and global operations with the latter being divided into the goals and benefits for management, R&D, and field service.

Production flow monitoring: optimize flow, eliminate waste, and avoid unnecessary work in process inventory. Remote equipment management, including setting

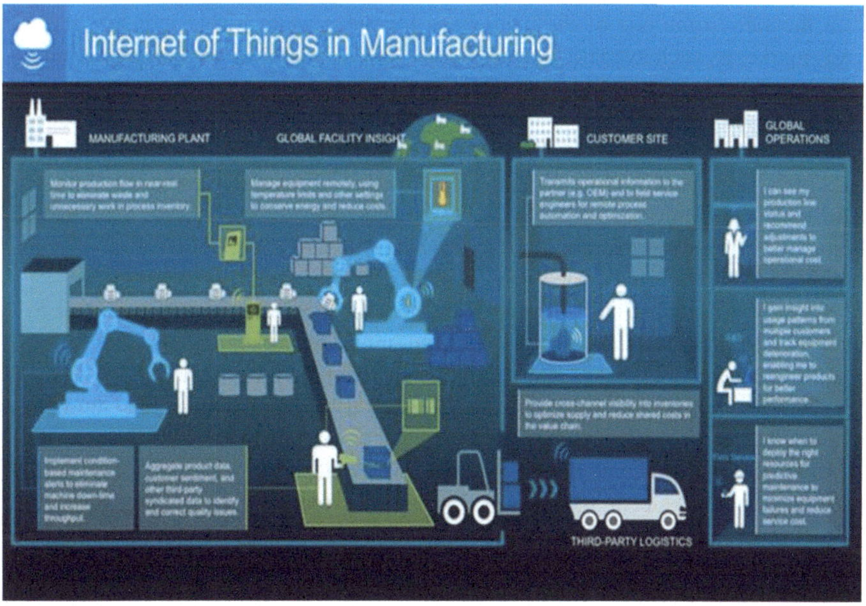

Fig. 3 Internet of things in manufacturing

specific limits and parameters to save energy and costs. Condition-based maintenance alerts: optimize machine availability, minimize interruption, and increase throughput.

The usage of various data (product, customer sentiment, and more) as a driver of quality monitoring and enhancement in function of outcomes.

Outside of the plant and actual manufacturing environment (insights and production) we see the transmission of operational information for those field service engineers on customer sites, the overall value and supply chain perspective, third-party logistics, and the global operations dimension.

2.2 The IIoT in Connected Logistics and Transportation

Transportation is the second largest market from an Internet of Things spending perspective. Transportation and logistics (T&L) industry can use the IoT in communication and monitoring system to increase profit.

The transportation market reached an IoT spend of $78 billion and is poised to continue to grow rapidly, just as is the case for the IoT manufacturing market. The main use case in transportation is freight monitoring, good for a large majority of overall transportation IoT spend with a total of $55.9 billion and remaining a key driver in the market until 2020 [9].

If we look at the overall IIoT evolutions in transportation and logistics, we see the growing emergence of a digital supply chain and connected logistics reality, which is at the same time one of the challenges for the manufacturing industry and the T&L market as such as many players do not have a digital strategy in place and are urged to speed up their digital transformation efforts.

As defined by Technavio, there are four pillars of logistic system: IT security, communication systems, supply chain monitoring systems, and vehicle/transport tracking. Along with the cloud and analytics, the Industrial Internet of Things is a driver in the connected logistics landscape and freight monitoring leads the pack. Figure 4 shows the four pillars of logistics: IT security, communication systems, supply chain monitoring system, and vehicle tracking.

Smart supply chain management (Logistics 4.0) is also data-intensive and IoT-intensive given with a focus on (semi-)autonomous decisions.

Logistics, transportation, and warehousing (supply chain management *overall)* are the first use of the concepts of connected devices which can sense and communicate.

With RFID and other connected logistics possibilities, the logistics market tries to build competitive benefits in a world where speed matters more than ever.

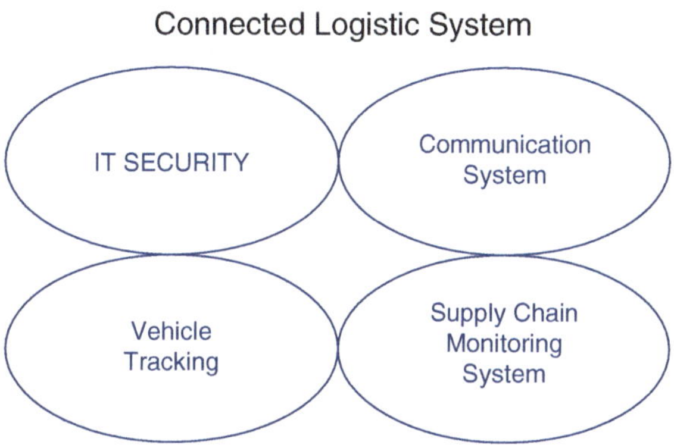

Fig. 4 Four pillars of connected logistic system [9]

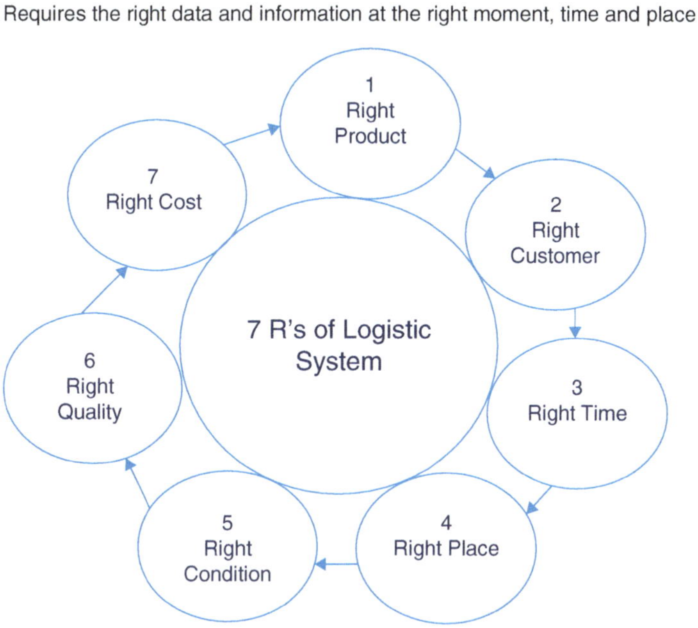

Fig. 5 Seven R's of logistic system

As getting the right product to the right customer at the right time, right place and right condition in the right quantity and at the right cost *(the famous 7R's of logistics)* is increasingly a challenge, the logistics industry keeps using connected logistics solutions, strongly powered by the Industrial Internet of Things, where it represents the second largest industry from a spending perspective, only preceded by manufacturing. Figure 5 [9] shows the seven R's of Logistics System.

2.3 IIoT in Energy and Utilities

Oil and gas, smart grid, and plenty of other evolutions and use cases in the energy and utilities market overall are also a main part of the Industrial Internet of Things market.

The Industrial Internet of Things plays a key role in the overall digital transformation towards a digital supply chain in many parts and value chain components of the large ecosystem, which obviously also touches retail/consumer-facing aspects.

However, from the sheer Industrial Internet perspective, smart grids are key in supply and network transmission/distribution. Others include plant effectiveness, maintenance, and data-driven opportunities as a result of smart grids and IoT-enabled operations and services. Figure 6 shows the different components of energy and utility industry which are customer, business, technology, data, competition, innovation, resources, and regularity.

2.4 IIoT in Other Industries

Maintenance and services enabled by the IIoT are two key areas in virtually all Industrial Internet industries.

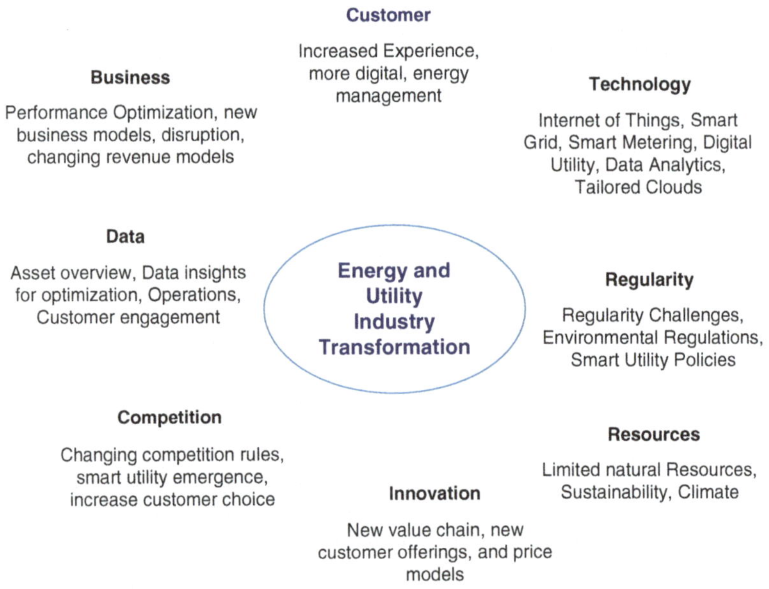

Fig. 6 IoT in energy and utility

Predictive maintenance, data-enabled services and remote possibilities in several areas, from service to control and optimization of operations, also come back in many other IIoT use cases across industries such as healthcare (remote health monitoring, equipment maintenance, etc.), aviation, robotics and cobots, oil and gas, mining, metals, and more.

As mentioned, depending on the industry body, the broader context of IIoT is also often used for use cases in areas such as agriculture, smart cities, and so on.

3 IoT in Day-to-Day Life

IoT is entering into our day-to-day life. It is touching every aspect of our life like health, agriculture, smart homes, smart cities, energy consumption, etc. Following are some of the applications of IoT in our day-to-day life [10].

3.1 Tuning Your Car

As more machines speak to each other and systems integrate, you will no longer miss an oil change. Your truly "smart" car will preemptively reach out to your mechanic when it is time for the annual tune up or your tire pressure is running low, and by cross referencing your calendar, appointment suggestions will be delivered to you to confirm a time with one click.

3.2 Monitoring Your Health

When a prescription is running low, an appointment will be made with your physician through connected RX bottles. Doctors will be kept informed with how often and when their patients are taking their medicine and those with ongoing health issues will be able to have things such as blood pressure and sugar levels monitored remotely.

3.3 Energy Consumption

High-energy consumption household appliances will adjust based on dynamic price signals to lower your electric bill. Thermostats and lighting will learn your habits to create the optimal setting based on your daily life, such as turning to your ideal temperature just before you arrive home. These gadgets will also sense when no one is in the house and turn off automatically to reduce wastes and costs.

3.4 Driving and Traffic Jams

Driving will get a lot safer. Traffic lights will be able to adjust to real-time traffic conditions such as when an emergency vehicle is approaching. Road sensors will make changes to the speed limit based on weather and accidents, while also communicating directly to car dashboards about unsafe conditions (e.g., slow down. The turn in a quarter mile is icy).

3.5 Grocery Lists

Smart refrigerators will sense when you are running low on staples such as eggs or milk and will automatically populate your grocery list. Stores will push reminders to add items to your list when it predicts you about to run out based on your historical purchasing behavior and average buying trends. When you are walking through the store, reminders will get pushed to you to ensure you never have to make that dreaded second trip.

3.6 Our Morning Alarm

The traffic on your route to work and the weather will soon affect what time your alarm goes off. If there is an accident or road construction on your usual drive, your alarm will go off early and alternate routes will populate in your dashboard. Of course, your coffee machine will be in the loop to make sure you have your cup of joe for the road.

3.7 Monitoring Your Baby

Through their smartphones, parents will monitor their baby's breathing, temperature, and activity. Babies will don connected onesies that will send an alert when there is anything abnormal. Of course, the other babies in your life will also reap the benefits of connectivity. Pet monitoring systems will allow you to monitor their activity and behavior from afar, so you can see how well your potty training is working and how honest your dog walker really is.

3.8 What's on Your Body

Wearable tech has perhaps gotten the most attention in the Internet of Things chatter to date. Many products are now in their second or third generations, offering sleeker designs and more integration with different systems. From monitoring activity during workouts to sleeping patterns to hearing aids, the devices that we "wear" are becoming much more sophisticated, connecting to all of our social media accounts, and tracking much more quality and quantity data.

3.9 IoT in Agriculture

With the continuous increase in world's population, demand for food supply is extremely raised. Governments are helping farmers to use advanced techniques and research to increase food production. Smart farming is one of the fastest growing fields in IoT.

Farmers are using meaningful insights from the data to yield better return on investment. Sensing for soil moisture and nutrients, controlling water usage for plant growth, and determining custom fertilizer are some simple uses of IoT.

3.10 Smart Retail

The potential of IoT in the retail sector is enormous. IoT provides an opportunity to retailers to connect with the customers to enhance the in-store experience.

Smartphones will be the way for retailers to remain connected with their consumers even out of store. Interacting through smartphones and using beacon technology can help retailers serve their consumers better. They can also track consumer's path through a store and improve store layout and place premium products in high traffic areas.

3.11 IoT in Poultry and Farming

Livestock monitoring is about animal husbandry and cost saving. Using IoT applications to gather data about the health and well-being of the cattle, ranchers knowing early about the sick animal can pull out and help prevent large number of sick cattle. With the help of the collected data and ranchers can increase the poultry production.

4 Discussion

The Internet of Things is now not only limited to research purposes, but it is also getting proliferated in every aspect of human life and in the industry. This technology is continuously evolving, and new sensors and applications of it are coming on the market. The technology enables huge data collection, automation, operations and much more through smart devices and powerful enabling technologies. The key feature of the IoT includes artificial intelligence, smart devices, connectivity, sensors and active engagement. The IoT never works in isolation; it gets integrated with Big Data technology, machine learning, artificial intelligence, etc. to produce the results as per the requirements. The IoT has various advantages such as it improves customer engagement and technology optimization, reduces waste and enhances data collection; at the same time it needs to take care of issues such as security, privacy, complexity, flexibility of the IoT-based applications and compliance with other technologies.

5 Conclusion

The Internet of Things may represent the next big leap ahead in each and every aspect of our life. The possibility of seamlessly merging the real and the virtual world, through the massive deployment of embedded devices, opens up new exciting directions for both research and business.

In this chapter, we provided an overview of the key applications of the IoT in industry and in our day-to-day life. The chapter describes the most relevant applications in the field of manufacturing, energy, logistics, transport, etc. With the help of IoT, the day-to-day actions are also getting easy and people are more dependent on it. As the IoT continues to grow, the dependence of industry and people on it will continue to grow.

References

1. Retrieved from https://www.forbes.com/sites/louiscolumbus/2017/12/10/2017-roundup-of-internet-of-things-forecasts/#3ed98fe1480e
2. Retrieved from https://www.mckinsey.com/industries/semiconductors/our-insights/whats-new-with-the-internet-of-things
3. Retrieved from http://h30614.www3.hp.com/collateral/Barcelona2013/presentations/IT3112.pdf
4. Retrieved from https://www.ibm.com/blogs/internet-of-things/what-is-the-iot/
5. Retrieved from http://www.gsma.com/newsroom/wp-content/uploads/15625-Connected-Living-Report.pdf
6. Retrieved from https://automationforum.in/t/what-is-iiot-benifits-and-applications/4153
7. Retrieved from http://iot-analytics.com/wp/wp

8. *Digital transformation in the manufacturing industry: Challenges and accelerators*. Retrieved from https://www.i-scoop.eu/digital-transformation/digital-transformation-manufacturing
9. *Digital transformation in transportation and logistics*. Retrieved from https://www.i-scoop.eu/digital-transformation/transportation-logistics-supply-chain-management/connected-logistics-2017-2020
10. Retrieved from https://us.norton.com/internetsecurity-iot-5-predictions-for-the-future-of-iot.html

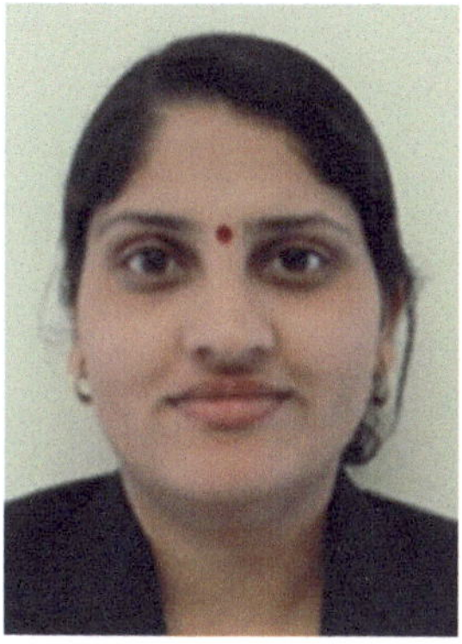

Kavita Suryawanshi is presently working as HOD (MCA) with Dr. D. Y. Patil Institute of Master of Computer Applications and Management. She has completed MCA and awarded Ph.D. She secured 13th Merit Rank holder in Maharashtra Higher Secondary State Board Examination. She has awarded with National Merit Scholarship for higher education. She holds a University Topper during her graduation from North Maharashtra University, Jalgaon.

She also won Best Paper Award at National Conference ETIT. She has got total 14 years of experience. She has presented many papers at international conference organized by renowned professional bodies like IEEE, CSI, Springer, Elsevier, etc. and also published several research papers in national and international journals of high repute. She believes in consistent untiring efforts towards goal achievement.

Manasi Pandharkar is currently working as an Assistant Professor in D. Y. Patil Institute of MCA and Management, Akurdi, Pune. She has completed MCA from S.N.D.T. University and currently pursuing Ph.D. in Recommendation System from Mumbai University. She has more than 10 years of teaching experience and areas of specializations are database management system, data mining and warehousing, advanced java, Internet of Things, machine learning, R Programming, etc. She has written many research papers in national and international conferences.

An Overview of Energy Harvesting Techniques for Future Internet of Things Applications

M. Saravanan, J. Ajayan, Sathish R. Jondhale, and P. Mohankumar

1 Introduction

Energy harvesting is a promising technology used to overcome the drawback of low lifetime provided by traditional battery technologies [1–6]. EH is also called as energy scavenging. Energy harvesting is defined as the process by which energy is extracted from energy sources like sunlight, wind, heat, ocean waves, electromagnetic waves, geothermal energy, and mechanical vibrations and their storage. Energy harvesting is a technique that can be used to drive energy from the environment. Energy crisis and global warming are the two most critical issues to be addressed in the near future. Utilization of green energy and renewable sources is essential to meet the global energy needs. But one drawback of these green energy and renewable sources is that their availability depends on weather conditions. Energy-harvesting devices can be used to support a wide variety of self-powered portable wireless electronic applications [7–15]. Nanogenerators are the heart of all the energy-harvesting system which is used to convert the external sources of energy such as sunlight, wind, mechanical vibrations, chemical, heat, and electromagnetic waves into useful electric energy [15]. Energy-harvesting technologies can be classified as follows:

M. Saravanan · J. Ajayan (✉)
Department of Electronics and Communication Engineering, SNS College of Technology, Coimbatore, India

S. R. Jondhale
Department of Electronics and Telecommunication Engineering, Amrutvahini College of Engineering, Sangamner, Maharashtra, India

P. Mohankumar
Department of Mechatronics Engineering, SNS College of Technology, Coimbatore, India

G. R. Kanagachidambaresan et al. (eds.), *Internet of Things in Smart Technologies for Sustainable Urban Development*, EAI/Springer Innovations in Communication and Computing, https://doi.org/10.1007/978-3-030-34328-6_7

a. Thermoelectric (TEEH)
b. Pyroelectric (PyEEH)
c. Photovoltaic (PVEH)
d. Piezoelectric (PEEH)
e. Electrostatic (ESEH)
f. RF and electromagnetic (RF&EM EH)
g. Wind (WEH)
h. Vibration (VEH)
i. Biochemical (BCEH)

Nanogenerators are of two types, and they are single nanogenerators and hybrid nanogenerators. Single nanogenerators convert one form of energy into electrical energy. Example: A single nanogenerator can be used to convert sunlight into electricity. On the other hand, a hybrid nanogenerator converts multiple forms of energy into electrical energy. A hybrid nanogenerator can take sunlight, heat, and mechanical vibrations as their input energy.

2 Thermoelectric Energy Harvesting (TEEH)

Applications such as healthcare, smart city, production plants, machines, and vehicles utilize wireless sensor networks for monitoring. This is because WSNs do not need wires. Batteries are widely used to power these wireless sensor. The main disadvantage of using batteries for powering these sensors is that the battery must be replaced or recharged periodically. In order to alleviate these difficulty, EH can be adopted. In thermoelectric energy harvesting (TEEH), a thermoelectric generator can be used to convert thermal energy into electricity [16–54]. Telluride alloys are the popular thermoelectric materials used in thermoelectric generators [16].

The schematic of a heat storage TEEH device reported by Michail E Kiziroglou [24] is shown in Fig. 1. In Fig. 1 the heat storage unit consists of a phase change material (PCM) embedded in a metallic container. Thermal bridges are used to enhance the thermal properties of PCM-TEG. The diagram of a typical thermoelectric generator (TEG) is shown in Fig. 2. $BiTe_3$ and $CoSb_3$ are the popular thermoelectric materials used in TEGs. A good thermoelectric material should have low thermal conductance and high power factor [26]. The insulating layer is used to prevent the leakage of heat to the environment. The heat-sink attached to the TEG helps to improve the thermal contact with the surroundings. A substantial variation in environmental temperature leads to the flow of heat in and out of the heat storage unit through the TEG resulting in the production of electricity [24]. The diagram of a thermoelectric microconverter is used to convert heat into electricity as shown in Fig. 3.

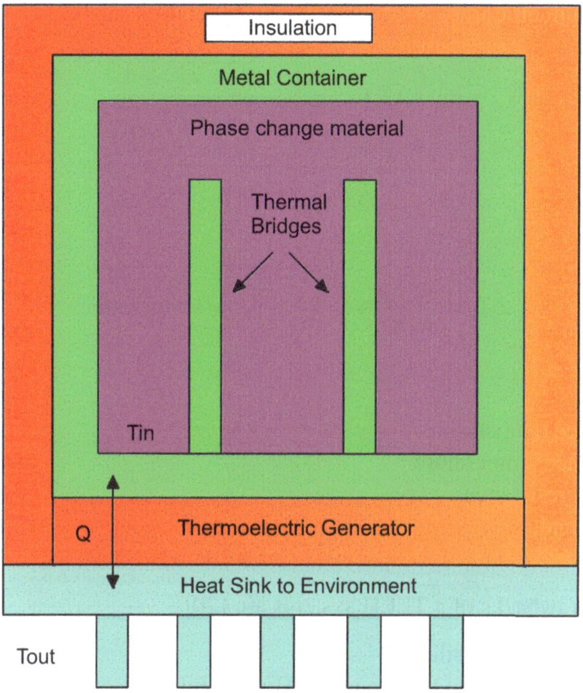

Fig. 1 Schematic of a heat storage TEEH device [24]

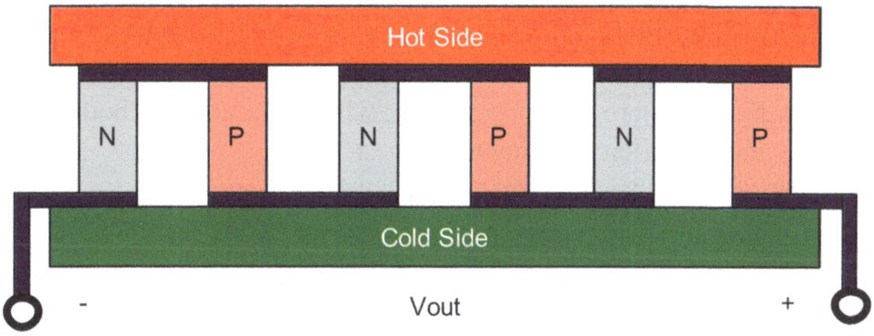

Fig. 2 Schematic of typical thermoelectric generator [29]

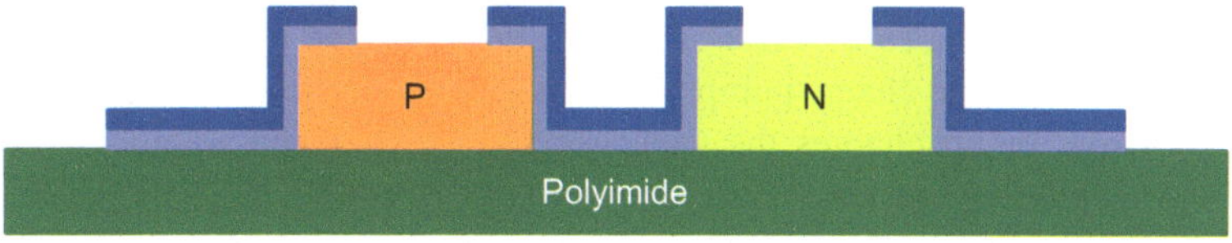

Fig. 3 The diagram of a thermoelectric microconverter [16]

The maximum efficiency of thermoelectric generators can be computed as [24]

$$\eta_{\text{TEG}}\,(\Delta T) = \frac{\Delta T}{T_{\text{h}}} \frac{\sqrt{1+ZT}-1}{\sqrt{1+ZT}+\frac{T_{\text{C}}}{T_{\text{h}}}} \tag{1}$$

$$\Delta T = T_{\text{h}} - T_{\text{C}} \tag{2}$$

$$ZT = \frac{a^2 T}{R_{\text{e}} K} \approx 0.8 \quad \text{at room temperature} \tag{3}$$

where

T_{h} = hot surface temperature
T_{c} = cold surface temperature
R_{e} = electrical resistance
K = thermal conductance
a = seeback coefficient

The main drawbacks of a TEEH systems are [20]

(a) No self-start-up capability
(b) Maximum power point tracking (MPPT) capability
(c) Poor energy transfer efficiency

These drawbacks can be eliminated by using transformer-reuse self-start-up boost converter with MPPT control [20, 21]. Wearable sensor can be powered by TEEH devices that can convert human body temperature into electricity [22]. The schematic of a hybrid micro power generator which can connect both optical and thermal energy into electricity is shown in Fig. 4.

The applications of thermoelectric energy harvesting are

1. To power the internal start-up circuit for pacemakers [29]
2. To power wearable and implantable medical devices [39]
3. To power wireless sensors in nuclear power plants [36]

The architecture of a TEEH system is illustrated in Fig. 5. The TEG output voltage acts as the input to the thermoelectric energy-harvesting system. A low-power oscillator produces the desired clock signal for the charge pump system. The second input to the charge pump system is the thermoelectric output generator voltage. Charge pump charges the capacitor C_{PST}. When the voltage across the capacitor exceeds the threshold level, the output of first comparator sets activates the start-up boost converter. The start-up boost converter generates the required clock pulses, for the steady-state boost converter, while the output voltage (Vout) does not exceed the present value. When the output voltage is equal to the present value of the second comparator, the steady-state boost converter generate clock signals. The

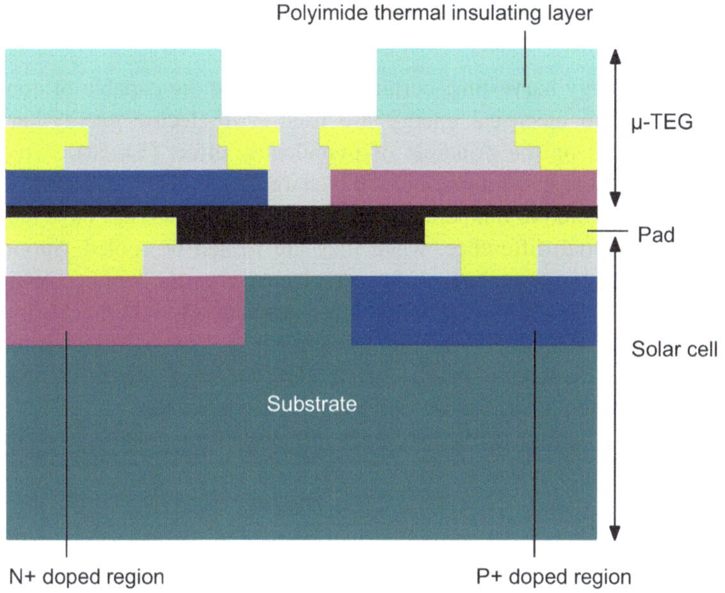

Fig. 4 Structure of a hybrid micropower generator [33]

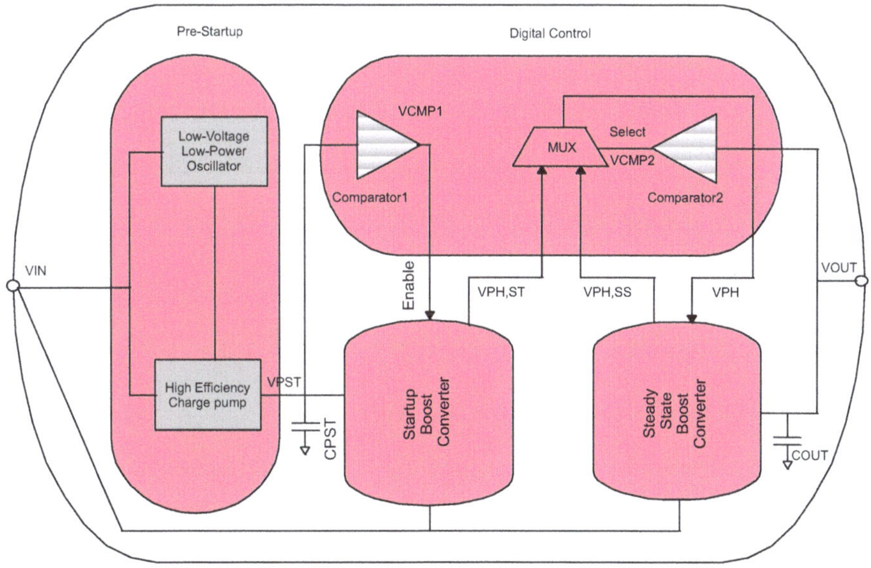

Fig. 5 The architecture of a TEEH system [29]

multiplexer selects the required signals for the steady-state boost converter based on the output voltage of the second comparator.

3 Pyroelectric Energy Harvesting (PyEEH)

In pyroelectric energy harvesting, certain materials that are capable of converting thermal energy into electrical energy are used. Pyroelectric energy-harvesting devices work based on the principle of pyroelectric effect [54–58]. Pyroelectric effect is defined as the generation of electric charge at the surface of a pyroelectric material when subjected to temperature changes. Certain materials have the ability to produce a potential difference, when they are heated or cooled. Spontaneous polarization is the unique feature of the pyroelectric crystals. Unlike TEGs, pyroelectric materials do not require a temperature gradient. $(1 - x)$ P_b $(Mg_{1/3}\ Nb_{2/3})$ $-xPbTiO_3$ (PMN-PT) single crystal is considered as the best pyroelectric materials [55]. PZT (lead zirconate titanate), PZT/PVDF (polyvinylidene flouride)-HFP composite, etc. are the other examples of pyroelectric materials.

The equations used to describe pyroelectric materials are given below [56].

$$dD = \varepsilon_{33}^{\theta} dE + p d\theta \tag{4}$$

$$C = \left(\frac{du}{d\theta}\right)_E \tag{5}$$

where

$D =$ dielectric displacement
$E =$ dielectric field
$\theta =$ temperature
$P =$ pyroelectric coefficient
$u =$ internal energy
$\varepsilon_{33}^{\theta} =$ dielectric permittivity
$C =$ heat capacity

$$\text{The energy harvested } (W_{\text{harvested}}) = \oint E d\text{P} \tag{6}$$

$$E = \frac{P^* \Delta T}{\varepsilon_r \varepsilon_0} \tag{7}$$

The energy received from the hot temperature medium is given by [56]

$$\theta_{\text{hot}} = C\,(\theta_2 - \theta_1) + \int_0^{E_{\text{m}}} P\theta_2 dE \tag{8}$$

The instantaneous pyroelectric current under short circuit conditions can be calculated as [58]

$$I = P^* A \frac{dT}{dt} \tag{9}$$

The efficiency is obtained as [56]

$$\text{Efficiency } (\eta) = \frac{W_{\text{harvested}}}{Q_{\text{hot}}} \tag{10}$$

where

P^* = pyroelectric coefficient
A = surface area of the material

4 Photovoltaic Energy Harvesting (PVEH)

Factors like type of applications, characteristic of the photovoltaic cells, power supply required, and capacity of the storage element are to be considered while designing a PVEH system [59–72]. Under typical indoor lighting conditions, a single crystal photovoltaic cell has an efficiency of 1–3%, whereas an amorphous photovoltaic cell can provide an efficiency of 3–7%. But amorphous photovoltaic cells are highly expensive than single crystal photovoltaic cells. Photovoltaic cells are the heart of PVEH system. The configuration of an indoor PVEH system is shown in Fig. 6. The equivalent circuit of a PV cell as shown in Fig. 7.

The current output of a PV cell can be computed as [71]

$$I = I_{\text{PG}} - I_0 \left(e^{\frac{V + R_S I}{V_t}} - 1 \right) - \frac{V + R_S I}{R_p} \tag{11}$$

where

I_{PG} = photo generated current

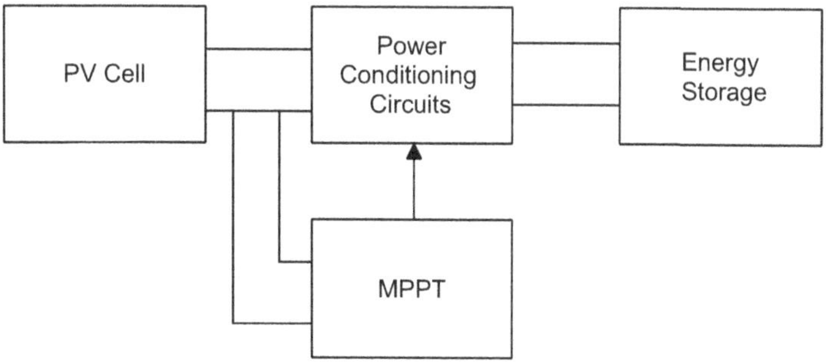

Fig. 6 The configuration of an indoor PVEH system

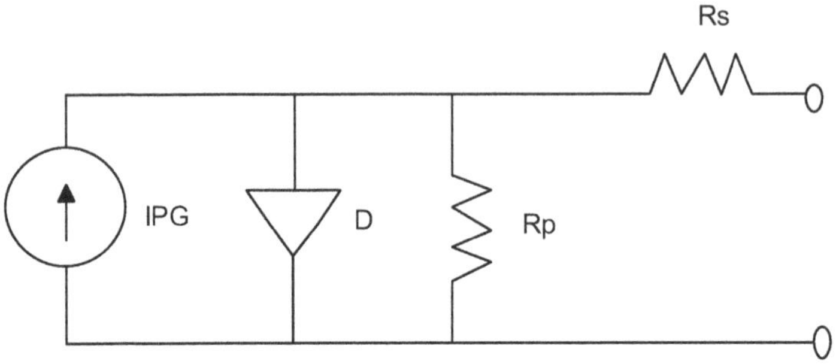

Fig. 7 Equivalent circuit of a PV cell [71, 72]

I_0 = dark current
R_s = series resistor
R_p = shunt resistor
V_t = junction terminal voltage

Battery/super capacitors can be used as energy storage elements. A photovoltaic cell directly converts solar energy into electric energy. PVEH systems find applications in autonomous systems, power plant systems, and WSNs. The main limitation of a PV cell is its poor efficiency (10–40%). Also the output voltage of a PV cell varies with changing environmental conditions. For enhancing the efficiency of PVEH system, a maximum power point tracking (MPPT) unit is required. Photovoltaic cells made of III-V compound semiconductors like GaAs provide large conversion efficiency than silicon-based PV cells due to their lower bandgap.

5 Piezoelectric Energy Harvesting (PEEH)

The demand for PEEH device is increasing day by day to the tremendous growth in low-power and wearable electronic devices [73–97]. Piezoelectric transducers can be used to extract electricity from the environmental energy sources. Piezoelectric harvesting devices uses piezoelectric materials like lead zirconium titanate (PZT) and polyvinylidene flouride-poly triflouro-ethylene (PVDF-PTrFE), ZnO, lead-magnesium-niobate-lead titanate (PMN-PT), etc. for converting mechanical stress into electrical energy. High power density, excellent scalability, and high output voltage are the main advantages of piezoelectric harvesters. When a mechanical stress is applied to a piezoelectric harvester, an emf will be induced in the piezoelectric material present inside the piezoelectric harvester. These are three commonly used methods to extract energy from the piezoelectric harvester.

1. A full wave bridge rectifier
2. Switch-only rectifier
3. Parallel synchronized switch harvesting on inductor

One of the main limitations of piezoelectric energy harvesting device is its high output voltage in conjunction with very low output current. Traditional piezoelectric harvesters can generate voltage output up to 100 V with an output current in the order of micro amperes. Therefore, transformers are required to reduce the voltage to the required level for applications which increase the size and cost of the devices. There are different types of piezoelectric generators and they are [90]:

1. Piezoelectric generators powered by human
2. Piezoelectric generators using vibration-based cantilever

6 Electromagnetic Energy Harvesting

Electromagnetic generators are used in electromagnetic energy harvesting devices to convert electromagnetic energy into electricity based on the principle of Faraday's law of electromagnetism [98–119]. The electromotive force for the microgenerator can be expressed as [119]

$$V_i = n\frac{d\phi}{dt} = n\left(\frac{\partial\phi}{\partial z}\left(\frac{\partial z}{\partial t}\right)\right) \tag{12}$$

where

ϕ = flux linkage per turn
n = number of windings in the coil
Z = internal displacement of the microgenerator

When the generator is connected to the load (R_{Load}), the current flowing through the load is given by

$$i = \frac{V_i}{R_{\text{Load}}} \tag{13}$$

The model of an electromagnetic energy harvester is shown in Fig. 8.

The electromagnetic energy harvester consists of a fixed magnet fixed at the bottom of the cap, a movable magnet in a cylindrical tube and a coil. The magnets generate repulsing forces. Electricity is produced in the coil based on magnetic induction. The emf generated in the coil as a function of time can be expressed as

$$E(t) = \frac{d\phi}{dt} \tag{14}$$

Using electromagnetic energy harvesting devices, the incident electromagnetic radiation can be converted into either AC or DC power [116–144]. The conversion efficiency can be expressed as

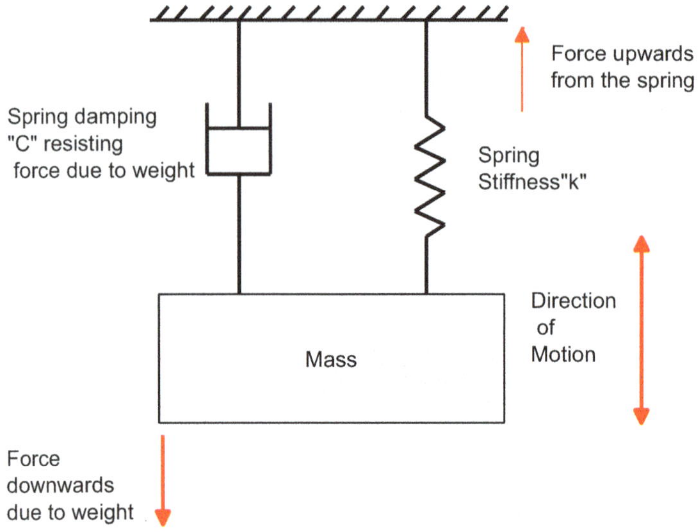

Fig. 8 The model of an electromagnetic energy harvester [124]

$$\eta_{\text{Rad-ac}} = \frac{P_{\text{out}}}{P_{\text{in}}} \tag{15}$$

where

$\eta_{\text{Rad - ac}}$ = radiation to alternating current (AC) conversion efficiency
P_{out} = average AC power received by the harvester
P_{in} = average incident power

$$\eta_{\text{Rad-dc}} = \eta_{\text{Rad-ac}} \times \eta_{\text{ac-dc}} \tag{16}$$

where

$\eta_{\text{Rad - dc}}$ = radiation to direct current (DC) conversion efficiency
$\eta_{\text{ac - dc}}$ = AC to DC conversion efficiency

The electromagnetic waves are available from different wireless communication technologies like Wi-Fi, WLAN, cellular, radio, and TV. A rectifier and a regulator circuit convert the AC power into DC power. The schematic of an RF/electromagnetic energy-harvesting system is shown in Fig. 9.

The microstrip antennas receive the RF signals coming from Wi-Fi, DTV, or UMT, and matching circuits are used to match the impedance of antenna and harvesting circuits. An RF to DC rectifier converts the AC signal into DC signal, and the voltage multiplier is used to increase the amplitude of the voltage, and the enhanced energy is finally stored in an energy storage device.

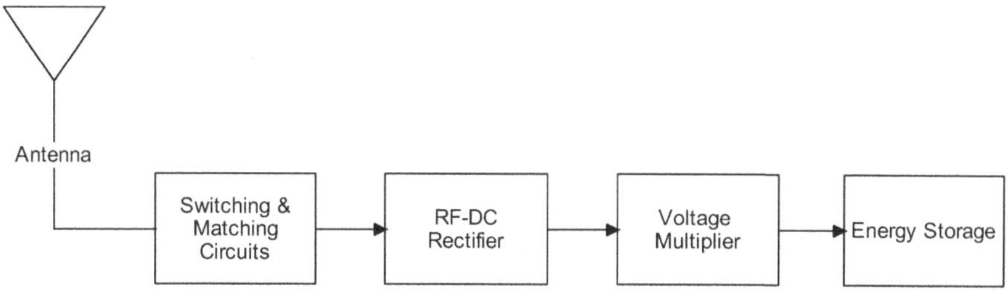

Fig. 9 RF/electromagnetic energy-harvesting system

7 Wind Energy Harvesting (WEH)

An air of mass "*m*" flowing with a velocity of V_{wind} has a kinetic energy E_{wind} which can be calculated as [150]

$$E_{\text{wind}} = \frac{1}{2} m V_{\text{wind}}^2 \tag{17}$$

The power of airflow available at the vertical cylinder with disc area "*A*" can be computed as [150]

$$P = \frac{1}{2} \rho \pi R^2 V_{\text{wind}}^3 \tag{18}$$

where

$A = \pi R^2$ is the area of the disc
$P =$ air density

In a WEH system, wind turbine generators are used to convert the mechanical power into electricity [145–153]. A wind turbine generator has mainly two parts: a stator and a rotor. Wind turbine generator works on the principle of electromagnetic induction. The blades of the wind turbine generator capture the mechanical energy of wind and transfer the power to the rotor. The movement of rotor creates a time-varying electromagnetic field that induces an electromotive force in across the stator winding. Wind turbine generators produce AC power. The output voltage obtained from a wind turbine generator can be expressed as [150]

$$V_{\text{turbine}} = V_{\text{s}} \times \frac{R_{\text{L}}}{R_{\text{L}} + jX_{\text{S}} + R_{\text{A}}} \tag{19}$$

where

$V_{\text{s}} =$ magnitude of electromagnetic force induced in the stator
$X_{\text{s}} =$ magnitude of generator reactance

R_L = load resistance
R_A = winding resistance

Wind energy harvesting system can be used to develop self-powered WSNs.

8 Vibrational Energy Harvesting

Piezoelectric energy-harvesting devices have been considered as the most attractive choice for replacing traditional battery technologies for low-power electronic devices in WSNs, implantable devices, vehicles, and health monitoring systems [154–170]. A piezoelectric energy-harvesting device can only convert vibration energy from one direction into electricity. A change in the direction of vibration leads to the degradation of piezoelectric energy harvesting due to the inability of capturing the vibrational energy. To overcome these limitations, vibrational energy-harvesting devices were introduced [169]. A vibrational energy harvester converts vibrational energy into electrical energy using the principle of magnetic inductance [163]. Base, cantilever beams, a magnet, a magnetic material, and a coil are the parts of a common vibrational energy-harvesting device. When a vibrational energy-harvesting device is placed on a vibrating object like motor or engine, the vibrations leads to the displacement of the magnets associated with the coil. This results in the vibration of magnetic flux associated with the coil that induces an emf in the coil [163]. Traditional linear resonant structure-based vibrational energy harvesters have very low operating efficiency and narrow bandwidth [160]. The bandwidth can be increased by adopting the techniques such as mechanical stopper, generator array, or bistable structures [160]. Vibration energy generators work based on electromagnetic or electrostatic or piezoelectric effects to convert vibration energy into electrical energy.

9 Summary

This chapter highlights the basic concepts of energy-harvesting systems and their applications. In recent years, the rapid growth of area-efficient low-power semiconductor technologies has enabled the implementation of low-power wireless sensors. Such wireless sensors find applications such as building automation, communication, and biomedical implants. All these devices were powered by traditional batteries in the past. Due to the low life span and need for replacement at regular periods of time, batteries are not preferred for such applications. In this scenario, EH devices can be used to extract electric energy from the surrounding energy sources like solar energy, heat energy, wind energy, and vibrational energy. No doubt that energy-harvesting devices can play a vital role in future IoT applications.

References

1. Dorey, R. A. (2014). Integrated powder-based thick films for thermoelectric, pyroelectric, and piezoelectric energy harvesting devices. *IEEE Sensors Journal, 14*, 2177–2184.
2. Liu, X., Huang, L., & Ravichandran, K. (2016). A highly efficient reconfigurable charge pump energy harvester with wide harvesting range and two-dimensional MPPT for Internet of Things. *IEEE Journal of Solid-State Circuits, 51*, 1302–1312.
3. Trapanese, M. (2008). Optimization of a sea wave energy harvesting electromagnetic device. *IEEE Transactions on Magnetics, 44*, 4365–4368.
4. Yoon, K.-S., Hong, S.-W., & Cho, G.-H. (2017). Double pile-up resonance energy harvesting circuit for piezoelectric and thermoelectric materials. *IEEE Journal of Solid-State Circuits, 53*, 1049–1060.
5. Lee, W.-K., Schubert, M. J., & Ooi, B. (2018). Multi-source energy harvesting and storage for floating wireless sensor network nodes with long range communication capability. *IEEE Transactions on Industry Applications, 54*, 2606–2615.
6. Khan, A. A., Mahmud, A., & Ban, D. (2019). Evolution from single to hybrid nanogenerator: A contemporary review on multimode energy harvesting for self-powered electronics. *IEEE Transactions on Nanotechnology, 18*, 21–36.
7. Aktakka, E. E., & Najafi, K. A. (2014). Micro inertial energy harvesting platform with self-supplied power management circuit for autonomous wireless sensor nodes. *IEEE Journal of Solid-State Circuits, 49*, 2017–2029.
8. Chiou, A. Y.-C., & Hsieh, C.-C. (2016). A 137 dB dynamic range and 0.32 V self-powered CMOS imager with energy harvesting pixels. *IEEE Journal of Solid-State Circuits, 51*, 2769–2776.
9. Yildiz, F. (2009). Potential ambient energy-harvesting sources and techniques. *Journal of Technology Studies, 35*, 40–48.
10. Gould, C., & Edwards, R. (2016). Review on micro-energy harvesting technologies. In *2016 51st International Universities Power Engineering Conference (UPEC)* (pp.1–5). IEEE.
11. Graf, C., Hitzbleck, J., & Feller, T. (2014). Dielectric elastomer–based energy harvesting: Material, generator design, and optimization. *Journal of Intelligent Material Systems and Structures, 25*, 951–966.
12. Roy, S., Podder, P., & Mallick, D. (2015). Nonlinear energy harvesting using electromagnetic transduction for wide bandwidth. *IEEE Magnetics Letters, 7*, 1–4.
13. Ma, K., Li, J., & Li, X. (2018). IAA: Incidental approximate architectures for extremely energy-constrained energy harvesting scenarios using IoT nonvolatile processors. *IEEE Micro, 38*, 11–19.
14. Joon Kim, K., Cottone, F., & Goyal, S. (2010). Energy scavenging for energy efficiency in networks and applications. *Bell Labs Technical Journal, 15*, 7–29.
15. Chen, X., Shao, J., & Li, X. (2016). A flexible piezoelectric-pyroelectric hybrid nanogenerator based on P (VDF-TrFE) nanowire array. *IEEE Transactions on Nanotechnology, 15*, 295–302.
16. Carmo, J. P., Gonçalves, L. M., & Correia, J. H. (2009). Thermoelectric microconverter for energy harvesting systems. *IEEE Transactions on Industrial Electronics, 57*, 861–867.
17. Carlson, E. J., Strunz, K., & Otis, B. P. (2010). A 20 mV input boost converter with efficient digital control for thermoelectric energy harvesting. *IEEE Journal of Solid-State Circuits, 45*, 741–750.
18. Ramadass, Y. K., & Chandrakasan, A. P. (2010). A battery-less thermoelectric energy harvesting interface circuit with 35 mV startup voltage. *IEEE Journal of Solid-State Circuits, 46*, 333–341.
19. Ibragimov, A., Pleteit, H., & Pille, C. (2011). A thermoelectric energy harvester directly embedded into casted aluminum. *IEEE Electron Device Letters, 33*, 233–235.

20. Im, J.-P., Wang, S.-W., & Lee, K.-H. (2012). A 40mV transformer-reuse self-startup boost converter with MPPT control for thermoelectric energy harvesting. *IEEE Journal of Solid State Circuits, 47*, 3055–3067.
21. Kim, J., & Kim, C. (2012). A DC–DC boost converter with variation-tolerant MPPT technique and efficient ZCS circuit for thermoelectric energy harvesting applications. *IEEE Transactions on Power Electronics, 28*, 3827–3833.
22. Leonov, V. (2013). Thermoelectric energy harvesting of human body heat for wearable sensors. *IEEE Sensors Journal, 13*, 2284–2291.
23. Weng, P.-S., Tang, H.-Y., & Ku, P.-C. (2013). 50 mV-input batteryless boost converter for thermal energy harvesting. *IEEE Journal of Solid-State Circuits, 48*, 1031–1041.
24. Kiziroglou, M. E., Wright, S. W., & Toh, T. T. (2013). Design and fabrication of heat storage thermoelectric harvesting devices. *IEEE Transactions on Industrial Electronics, 61*, 302–309.
25. Verma, R., Bhattacharya, S., & Mahapatra, S. (2013). Thermoelectric performance of a single-layer graphene sheet for energy harvesting. *IEEE Transactions on Electron Devices, 60*, 2064–2070.
26. Choday, S. H., Lundstrom, M. S., & Roy, K. (2013). Prospects of thin-film thermoelectric devices for hot-spot cooling and on-chip energy harvesting. *IEEE Transactions on Components, Packaging and Manufacturing Technology, 3*, 2059–2067.
27. Carreon-Bautista, S., Eladawy, A., & Mohieldin, A. N. (2014). Boost converter with dynamic input impedance matching for energy harvesting with multi-array thermoelectric generators. *IEEE Transactions on Industrial Electronics, 61*, 5345–5353.
28. Chen, P.-H., & Fan, P. M.-Y. (2014). An 83.4% peak efficiency single-inductor multiple-output based adaptive gate biasing DC-DC converter for thermoelectric energy harvesting. *IEEE Transactions on Circuits and Systems I: Regular Papers, 62*, 405–412.
29. Ashraf, M., & Masoumi, N. (2015). A thermal energy harvesting power supply with an internal startup circuit for pacemakers. *IEEE Transactions on Very Large Scale Integration (VLSI) Systems, 24*, 26–37.
30. Shrivastava, A., Roberts, N. E., & Khan, O. U. (2015). A 10 mV-input boost converter with inductor peak current control and zero detection for thermoelectric and solar energy harvesting with 220 mV cold-start and $-$14.5 dBm, 915 MHz RF kick-start. *IEEE Journal of Solid-State Circuits, 50*, 1820–1832.
31. Dias, P. C., Morais, F. J. O., & De Morais França, M. B. (2015). Autonomous multisensor system powered by a solar thermoelectric energy harvester with ultralow-power management circuit. *IEEE Transactions on Instrumentation and Measurement, 64*, 2918–2925.
32. Petucco, A., Saggini, S., & Corradini, L. (2016). Analysis of power processing architectures for thermoelectric energy harvesting. *IEEE Journal of Emerging and Selected Topics in Power Electronics, 4*, 1036–1049.
33. Yan, J., Liao, X., & Ji, S. (2019). A novel multi-source micro power generator for harvesting thermal and optical energy. *IEEE Electron Device Letters, 40*, 349–352.
34. Qian, Y., Lu, D., & He, J. (2018). An on-chip transformer-based self-startup hybrid SIDITO converter for thermoelectric energy harvesting. *IEEE Transactions on Circuits and Systems II: Express Briefs, 65*, 1673–1677.
35. Katic, J., Rodriguez, S., & Rusu, A. (2016). A dual-output thermoelectric energy harvesting interface with 86.6% peak efficiency at 30$\mu {\text {W}} $ and total control power of 160 nW. *IEEE Journal of Solid-State Circuits, 51*, 928–1937.
36. Chen, J., Klein, J., & Wu, Y. (2016). A thermoelectric energy harvesting system for powering wireless sensors in nuclear power plants. *IEEE Transactions on Nuclear Science, 63*, 2738–2746.
37. Qian, Y., Zhang, H., & Chen, Y. (2016). A SIDIDO DC–DC converter with dual-mode and programmable-capacitor-array MPPT control for thermoelectric energy harvesting. *IEEE Transactions on Circuits and Systems II: Express Briefs, 64*, 952–956.
38. Alhawari, M., Mohammad, B., & Saleh, H. (2016). An efficient polarity detection technique for thermoelectric harvester in L-based converters. *IEEE Transactions on Circuits and Systems I: Regular Papers, 64*, 705–716.

39. Das, A., Gao, Y., & Kim, T. T.-H. (2016). A 220-mV power-on-reset based self-starter with 2-nW quiescent power for thermoelectric energy harvesting systems. *IEEE Transactions on Circuits and Systems I: Regular Papers, 64*, 217–226.
40. Rozgić, D., & Marković, D. (2017). A miniaturized 0.78-mW/cm^2 autonomous thermoelectric energy-harvesting platform for biomedical sensors. *IEEE Transactions on Biomedical Circuits and Systems, 11*, 773–783.
41. Katic, J., Rodriguez, S., & Rusu, A. (2017). A high-efficiency energy harvesting interface for implanted biofuel cell and thermal harvesters. *IEEE Transactions on Power Electronics, 33*, 4125–4134.
42. Yan, J., Liao, X., & Yan, D. (2018). Review of micro thermoelectric generator. *Journal of Microelectromechanical Systems, 27*, 1–18.
43. Kim, J., Baek, D., & Ding, C. (2018). Dynamic reconfiguration of thermoelectric generators for vehicle radiators energy harvesting under location-dependent temperature variations. *IEEE Transactions on Very Large Scale Integration (VLSI) Systems, 26*, 1241–1253.
44. Haug, B. (2017). Wireless sensor nodes can be powered by temperature gradients; No batteries needed: Harvesting energy from thermoelectric generators. *IEEE Power Electronics Magazine, 4*, 24–32.
45. Wan, Q., Teh, Y.-K., & Gao, Y. (2017). Analysis and design of a thermoelectric energy harvesting system with reconfigurable array of thermoelectric generators for IoT applications. *IEEE Transactions on Circuits and Systems I: Regular Papers, 64*, 2346–2358.
46. Kiziroglou, M. E., Becker, T., & Wright, S. W. (2017). Three-dimensional printed insulation for dynamic thermoelectric harvesters with encapsulated phase change materials. *IEEE Sensors Letters, 1*, 1–4.
47. Luo, Z., Zeng, L., & Lau, B. (2017). A sub-10 mV power converter with fully integrated self-start, MPPT, and ZCS control for thermoelectric energy harvesting. *IEEE Transactions on Circuits and Systems I: Regular Papers, 65*, 1744–1757.
48. Lei, P.-S., Chang, R. C.-H., & Chia, C.-H. (2014). Low start-up voltage dc–dc converter with negative voltage control for thermoelectric energy harvesting. *The Journal of Engineering, 2015*, 8–40.
49. Hou, L., Tan, S., & Zhang, Z. (2018). Thermal energy harvesting WSNs node for temperature monitoring in IIoT. *IEEE Access, 6*, 35243–35249.
50. Dezyani, M., Ghafoorifard, H., & Sheikhaei, S. (2018). A 60 mV input voltage, process tolerant start-up system for thermoelectric energy harvesting. *IEEE Transactions on Circuits and Systems I: Regular Papers, 65*, 3568–3577.
51. Lopera, J. M., Del Arco Rodríguez, H., & Pereira, J. M. P. (2019). Practical issues in the design of wireless sensors supplied by energy harvesting thermoelectric generators. *IEEE Transactions on Industry Applications, 55*, 996–1005.
52. Lim, B.-M., Seo, J.-I., & Lee, S.-G. (2018). A Colpitts oscillator-based self-starting boost converter for thermoelectric energy harvesting with 40-mV startup voltage and 75% maximum efficiency. *IEEE Journal of Solid-State Circuits, 53*, 3293–3302.
53. Verma, G., & Sharma, V. (2019). A novel thermoelectric energy harvester for wireless sensor network application. *IEEE Transactions on Industrial Electronics, 66*, 3530–3538.
54. Jo, S., Kim, M., & Kim, M. (2012). Flexible thermoelectric generator for human body heat energy harvesting. *Electronics Letters, 48*, 1015–1017.
55. Sebald, G., Lefeuvre, E., & Guyomar, D. (2008). Pyroelectric Energy Conversion: Optimization Principles. *IEEE Transactions on Ultrasonics, Ferroelectrics, and Frequency Control, 55*, 538–551.
56. Khodayari, A., Pruvost, S., & Sebald, G. (2009). Nonlinear pyroelectric energy harvesting from relaxor single crystals. *IEEE Transactions on Ultrasonics, Ferroelectrics, and Frequency Control, 56*, 693–699.
57. Mane, P., Xie, J., & Leang, K. K. (2011). Cyclic energy harvesting from pyroelectric materials. *IEEE Transactions on Ultrasonics, Ferroelectrics, and Frequency Control, 58*, 10–17.

58. Zabek, D., Taylor, J., & Bowen, C. R. (2016). Characterization and modeling of meshed electrodes on free standing polyvilylidene difluoride (PVDF) films for enhanced pyroelectric energy harvesting. *IEEE Transactions on Ultrasonics, Ferroelectrics, and Frequency Control, 63*, 1681–1689.
59. Wang, Y., Liu, Y., & Wang, C. (2015). Storage-less and converter-less photovoltaic energy harvesting with maximum power point tracking for internet of things. *IEEE Transactions on Computer-Aided Design of Integrated Circuits and Systems, 35*, 173–186.
60. Tar, B., & Çilingiroğlu, U. (2014). Nanowatt-scale power management for on-chip photovoltaic energy harvesting beacons. *IEEE Journal on Emerging and Selected Topics in Circuits and Systems, 4*, 284–291.
61. Teran, A. S., Wong, J., & Lim, W. (2015). AlGaAs photovoltaics for indoor energy harvesting in mm-scale wireless sensor nodes. *IEEE Transactions on Electron Device, 62*, 2170–2175.
62. Yu, G., Chew, K. W. R., & Sun, Z. C. (2015). A 400 nW single-inductor dual-input–tri-output DC–DC buck–boost converter with maximum power point tracking for indoor photovoltaic energy harvesting. *IEEE Journal of Solid-State Circuits, 50*, 2758–2772.
63. Brockett, T. J., Rajagopalan, H., & Laghumavarapu, R. B. (2013). Electromagnetic characterization of high absorption sub-wavelength optical nanostructure photovoltaics for solar energy harvesting. *IEEE Transactions on Antennas and Propagation, 61*, 1518–1527.
64. Yue, X., Kauer, M., & Bellanger, M. (2017). Development of an indoor photovoltaic energy harvesting module for autonomous sensors in building air quality applications. *IEEE Internet of Things Journal, 4*, 2092–2103.
65. Ahnood, A., & Nathan, A. (2012). Flat-panel compatible photovoltaic energy harvesting system. *Journal of Display Technology, 8*, 204–211.
66. Moon, E., Blaauw, D., & Phillips, J. D. (2017). Subcutaneous photovoltaic infrared energy harvesting for bio-implantable devices. *IEEE Transactions on Electron Devices, 64*, 2432–2437.
67. Prijić, A., Vračar, L., & Pavlović, Z. (2015). The effect of flat panel reflectors on photovoltaic energy harvesting in wireless sensor nodes under low illumination levels. *IEEE Sensors Journal, 15*, 7105–7111.
68. Ramli, M. Z., & Salam, Z. (2014). A simple energy recovery scheme to harvest the energy from shaded photovoltaic modules during partial shading. *IEEE Transactions on Power Electronics, 29*, 6458–6471.
69. Sajadian, S., & Ahmadi, R. (2017). Distributed maximum power point tracking using model predictive control for photovoltaic energy harvesting architectures based on cascaded power optimizers. *IEEE Journal of Photovoltaics, 7*, 849–857.
70. Mondal, S., & Paily, R. (2017). On-chip photovoltaic power harvesting system with low-overhead adaptive MPPT for IoT nodes. *IEEE Internet of Things Journal, 4*, 1624–1633.
71. Nasiri, A., Zabalawi, S. A., & Mandic, G. (2009). Indoor power harvesting using photovoltaic cells for low-power applications. *IEEE Transactions on Industrial Electronics, 56*, 4502–4509.
72. Kim, H., Kim, S., & Kwon, C.-K. (2012). An energy-efficient fast maximum power point tracking circuit in an 800-μW photovoltaic energy harvester. *IEEE Transactions on Power Electronics, 28*, 2927–2935.
73. Stein, A. L., & Hofmann, H. F. (2016). Autonomous wideband piezoelectric energy harvesting utilizing a resonant inverter. *IEEE Transactions on Power Electronics, 32*, 6178–6187.
74. Lee, S., & Youn, B. D. (2011). A new piezoelectric energy harvesting design concept: Multimodal energy harvesting skin. *IEEE Transactions on Ultrasonics, Ferroelectrics, and Frequency Control, 58*, 629–645.
75. Song, H.-C., Kumar, P., & Maurya, D. (2017). Ultra-low resonant piezoelectric MEMS energy harvester with high power density. *Journal of Microelectromechanical Systems, 26*, 1226–1234.
76. Shi, G., Xia, Y., & Wang, X. (2018). An efficient self-powered piezoelectric energy harvesting CMOS interface circuit based on synchronous charge extraction technique. *IEEE Transactions on Circuits and Systems I: Regular Papers, 65*, 804–817.

77. Shareef, A., Goh, W. L., & Narasimalu, S. (2019). A rectifier-less AC–DC interface circuit for ambient energy harvesting from low-voltage piezoelectric transducer array. *IEEE Transactions on Power Electronics, 34*, 1446–1457.
78. Do, X.-D., Nguyen, H.-H., & Han, S.-K. (2014). A self-powered high-efficiency rectifier with automatic resetting of transducer capacitance in piezoelectric energy harvesting systems. *IEEE Transactions on Very Large Scale Integration (VLSI) Systems, 23*, 444–453.
79. Wu, L., Do, X.-D., & Lee, S.-G. (2016). A self-powered and optimal SSHI circuit integrated with an active rectifier for piezoelectric energy harvesting. *IEEE Transactions on Circuits and Systems I: Regular Papers, 64*, 537–549.
80. Chen, Z., Law, M.-K., & Mak, P.-I. (2017). Fully integrated inductor-less flipping-capacitor rectifier for piezoelectric energy harvesting. *IEEE Journal of Solid-State Circuits, 52*, 3168–3180.
81. Dini, M., Romani, A., & Filippi, M. (2015). A nanopower synchronous charge extractor IC for low-voltage piezoelectric energy harvesting with residual charge inversion. *IEEE Transactions on Power Electronics, 31*, 1263–1274.
82. Romani, A., Filippi, M., & Tartagni, M. (2013). Micropower design of a fully autonomous energy harvesting circuit for arrays of piezoelectric transducers. *IEEE Transactions on Power Electronics, 29*, 729–739.
83. Rocha, J. G., Goncalves, L. M., & Rocha, P. (2009). Energy harvesting from piezoelectric materials fully integrated in footwear. *IEEE Transactions on Industrial Electronics, 57*, 813–819.
84. Reddy, A. R., Umapathy, M., & Ezhilarasi, D. (2015). Piezoelectric energy harvester with shape memory alloy actuator using solar energy. *IEEE Transactions on Sustainable Energy, 6*, 1409–1415.
85. Du, S., Amaratunga, G. A., & Seshia, A. A. (2019). A cold-startup SSHI rectifier for piezoelectric energy harvesters with increased open-circuit voltage. *IEEE Transactions on Power Electronics, 34*, 263–274.
86. Janphuang, P., Lockhart, R. A., & Isarakorn, D. (2014). Harvesting energy from a rotating gear using an AFM-like MEMS piezoelectric frequency up-converting energy harvester. *Journal of Microelectromechanical Systems, 24*, 742–754.
87. Miller, L. M., Elliott, A. D., & Mitcheson, P. D. (2016). Maximum performance of piezoelectric energy harvesters when coupled to interface circuits. *IEEE Sensors Journal, 16*, 4803–4815.
88. Kwon, D., & Rincón-Mora, G. A. (2010). A 2-$\mu $ m BiCMOS rectifier-free AC–DC piezoelectric energy harvester-charger IC. *IEEE Transactions on Biomedical Circuits and Systems, 4*, 400–409.
89. Priya, S. (2010). Criterion for material selection in design of bulk piezoelectric energy harvesters. *IEEE Transactions on Ultrasonics, Ferroelectrics, and Frequency Control, 57*, 610–2612.
90. Khaligh, A., Zeng, P., & Zheng, C. (2009). Kinetic energy harvesting using piezoelectric and electromagnetic technologies—State of the art. *IEEE Transactions on Industrial Electronics, 57*, 850–860.
91. Kodali, P., Krishna, A., & Varun, R. (2014). Segmented electrodes for piezoelectric energy harvesters. *IEEE Electron Device Letters, 35*, 85–487.
92. Kashyap, R., Lenka, T., & Baishya, S. (2015). A model for doubly clamped piezoelectric energy harvesters with segmented electrodes. *IEEE Electron Device Letters, 36*, 1369–1372.
93. Zhang, L., Oh, S. R., & Wong, T. C. (2013). Piezoelectric polymer multilayer on flexible substrate for energy harvesting. *IEEE Transactions on Ultrasonics, Ferroelectrics, and Frequency Control, 60*, 2013–2020.
94. Kamata, Y., Yoon, D. H., & Sasaki, T. (2016). Stacked piezoelectric energy harvesting device by printing process. *Micro & Nano Letters, 11*, 650–653.
95. Liang, J., & Liao, W.-H. (2011). Impedance modeling and analysis for piezoelectric energy harvesting systems. *IEEE/ASME Transactions on Mechatronics, 17*, 1145–1157.

96. Liang, J., & Liao, W.-H. (2011). Improved design and analysis of self-powered synchronized switch interface circuit for piezoelectric energy harvesting systems. *IEEE Transactions on Industrial Electronics, 59*, 1950–1960.
97. Lallart, M., Wu, Y.-C., & Guyomar, D. (2011). Switching delay effects on nonlinear piezoelectric energy harvesting techniques. *IEEE Transactions on Industrial Electronics, 59*, 464–472.
98. Kiziroglou, M. E., He, C., & Yeatman, E. M. (2008). Rolling rod electrostatic microgenerator. *IEEE Transactions on Industrial Electronics, 56*, 1101–1108.
99. Torres, E. O., & Rincón-Mora, G. A. (2008). Electrostatic energy-harvesting and battery-charging CMOS system prototype. *IEEE Transactions on Circuits and Systems I: Regular Papers, 56*, 1938–1948.
100. Tvedt, L. G. W., Nguyen, D. S., & Halvorsen, E. (2010). Nonlinear behavior of an electrostatic energy harvester under wide-and narrowband excitation. *Journal of Microelectromechanical Systems, 19*, 305–316.
101. Torres, E. O., & Rincón-Mora, G. A. (2010). A 0.7-$\mu $ m BiCMOS electrostatic energy-harvesting system IC. *IEEE Journal of Solid-State Circuits, 45*, 483–496.
102. Torres, E. O., & Rincón-Mora, G. A. (2010). Self-tuning electrostatic energy-harvester IC. *IEEE Transactions on Circuits and Systems II: Express Briefs, 57*, 808–812.
103. Kempitiya, A., Borca-Tasciuc, D.-A., & Hella, M. M. (2011). Analysis and optimization of asynchronously controlled electrostatic energy harvesters. *IEEE Transactions on Industrial Electronics, 59*, 456–463.
104. Mitcheson, P. D., & Green, T. C. (2012). Maximum effectiveness of electrostatic energy harvesters when coupled to interface circuits. *IEEE Transactions on Circuits and Systems I: Regular Papers, 59*, 3098–3111.
105. Kempitiya, A., Borca-Tasciuc, D.-A., & Hella, M. M. (2012). Low-power ASIC for microwatt electrostatic energy harvesters. *IEEE Transactions on Industrial Electronics, 60*, 5639–5647.
106. Yang, Z., Halvorsen, E., & Dong, T. (2013). Electrostatic energy harvester employing conductive droplet and thin-film electret. *Journal of Microelectromechanical Systems, 23*, 315–323.
107. Stanzione, S., Van Liempd, C., & Van Schaijk, R. (2013). A high voltage self-biased integrated DC-DC buck converter with fully analog MPPT algorithm for electrostatic energy harvesters. *IEEE Journal of Solid-State Circuits, 48*, 3002–3010.
108. Crovetto, A., Wang, F., & Hansen, O. (2014). Modeling and optimization of an electrostatic energy harvesting device. *Journal of Microelectromechanical Systems, 23*, 1141–1155.
109. De Queiroz, A. C. M. (2018). Analysis of electronic electrostatic generators. *IEEE Transactions on Circuits and Systems II: Express Briefs, 66*, 632–636.
110. Kaur, S., Halvorsen, E., & Soråsen, O. (2015). Characterization and modeling of nonlinearities in in-plane gap closing electrostatic energy harvester. *Journal of Microelectromechanical Systems, 24*, 2071–2082.
111. Karami, A., Galayko, D., & Basset, P. (2016). Series-parallel charge pump conditioning circuits for electrostatic kinetic energy harvesting. *IEEE Transactions on Circuits and Systems I: Regular Papers, 64*, 227–240.
112. Ahmed, S., & Kakkar, V. (2017). An electret-based angular electrostatic energy harvester for battery-less cardiac and neural implants. *IEEE Access, 5*, 19631–19643.
113. Truong, B. D., Le, C. P., & Halvorsen, E. (2017). Analysis of electrostatic energy harvesters electrically configured as Bennet's doublers. *IEEE Sensors Journal, 17*, 5180–5191.
114. Kim, D., Yu, S., & Kang, B.-G. (2015). Electrostatic energy harvester using magnetically actuated liquid dielectric layers. *Journal of Micro Electromechanical Systems, 24*, 516–518.
115. Lu, Y., Juillard, J., & Cottone, F. (2018). An impact-coupled MEMS electrostatic kinetic energy harvester and its predictive model taking nonlinear air damping effect into account. *Journal of Microelectromechanical Systems, 27*, 041–1053.
116. Bolt, R., Magno, M., & Burger, T. (2017). Kinetic AC/DC converter for electromagnetic energy harvesting in autonomous wearable devices. *IEEE Transactions on Circuits and Systems II: Express Briefs, 64*, 1422–1426.

117. Saha, C. R., O'donnell, T., & Loder, H. (2006). Optimization of an electromagnetic energy harvesting device. *IEEE Transactions on Magnetics, 42*, 3509–3511.
118. Liao, L.-D., Chao, P. C.-P., & Chen, J.-T. (2009). A miniaturized electromagnetic generator with planar coils and its energy harvest circuit. *IEEE Transactions on Magnetics, 45*, 4621–4627.
119. Dayal, R., Dwari, S., & Parsa, L. (2011). A new design for vibration-based electromagnetic energy harvesting systems using coil inductance of microgenerator. *IEEE Transactions on Industry Applications, 47*, 820–830.
120. Cannarella, J., Selvaggi, J., & Salon, S. (2011). Coupling factor between the magnetic and mechanical energy domains in electromagnetic power harvesting applications. *IEEE Transactions on Magnetics, 47*, 2076–2080.
121. Rahimi, A., Zorlu, Ö., & Muhtaroglu, A. (2011). Fully self-powered electromagnetic energy harvesting system with highly efficient dual rail output. *IEEE Sensors Journal, 12*, 2287–2298.
122. Maurath, D., Becker, P. F., & Spreemann, D. (2012). Efficient energy harvesting with electromagnetic energy transducers using active low-voltage rectification and maximum power point tracking. *IEEE Journal of Solid-State Circuits, 47*, 1369–1380.
123. Collado, A., & Georgiadis, A. (2013). Conformal hybrid solar and electromagnetic (EM) energy harvesting rectenna. *IEEE Transactions on Circuits and Systems I: Regular Papers, 60*, 2225–2234.
124. Yaşar, O., Uluşan, H., & Zorlu, Ö. (2018). Optimization of AA-battery sized electromagnetic energy harvesters: Reducing the resonance frequency using a non-magnetic inertial mass. *IEEE Sensors Journal, 18*, 4509–4516.
125. Erkmen, F., Almoneef, T. S., & Ramahi, O. M. (2018). Scalable electromagnetic energy harvesting using frequency-selective surfaces. *IEEE Transactions on Microwave Theory and Techniques, 66*, 2433–2441.
126. Kim, Y.-S. (2018). Induced voltage characteristics of back-iron effect for electromagnetic energy harvester using magnetic fluid. *IEEE Transactions on Applied Superconductivity, 28*, 1–4.
127. Ashoor, A. Z., Almoneef, T. S., & Ramahi, O. M. (2017). A planar dipole array surface for electromagnetic energy harvesting and wireless power transfer. *IEEE Transactions on Microwave Theory and Techniques, 66*, 1553–1560.
128. Wang, L., Todaria, P., & Pandey, A. (2016). An electromagnetic speed bump energy harvester and its interactions with vehicles. *IEEE/ASME Transactions on Mechatronics, 21*, 985–1994.
129. Zurbuchen, A., Haeberlin, A., & Bereuter, L. (2017). Endocardial energy harvesting by electromagnetic induction. *IEEE Transactions on Biomedical Engineering, 65*, 424–430.
130. Xie, L., & Cai, S. (2017). Electromagnetic energy-harvesting damper with multiple independently controlled transducers: On-demand damping and optimal energy regeneration. *IEEE/ASME Transactions on Mechatronics, 22*, 2705–2713.
131. Tang, Y., & Khaligh, A. (2015). A multiinput bridgeless resonant AC–DC converter for electromagnetic energy harvesting. *IEEE Transactions on Power Electronics, 31*, 2254–2263.
132. Paul, S., Bashir, S., & Chang, J. (2019). Design of a novel electromagnetic energy harvester with dual core for deicing device of transmission lines. *IEEE Transactions on Magnetics, 55*, 1–4.
133. Bito, J., Bahr, R., & Hester, J. G. (2017). A novel solar and electromagnetic energy harvesting system with a 3-D printed package for energy efficient Internet-of-Things wireless sensors. *IEEE Transactions on Microwave Theory and Techniques, 65*, 1831–1842.
134. Georgiadis, A., Andia, G. V., & Collado, A. (2010). Rectenna design and optimization using reciprocity theory and harmonic balance analysis for electromagnetic (EM) energy harvesting. *IEEE Antennas and Wireless Propagation Letters, 9*, 444–446.
135. Hemour, S., & Wu, K. (2014). Radio-frequency rectifier for electromagnetic energy harvesting: Development path and future outlook. *Proceedings of the IEEE, 102*, 1667–1691.

136. Jung, H.-J., Lee, S.-W., & Jang, D.-D. (2009). Feasibility study on a new energy harvesting electromagnetic device using aerodynamic instability. *IEEE Transactions on Magnetics, 45*, 4376–4379.
137. Reinisch, H., Gruber, S., & Unterassinger, H. (2011). An electro-magnetic energy harvesting system with 190 nW idle mode power consumption for a BAW based wireless sensor node. *IEEE Journal of Solid-State Circuits, 46*, 1728–1741.
138. Samad, F. A., Karim, M. F., & Paulose, V. (2015). A curved electromagnetic energy harvesting system for wearable electronics. *IEEE Sensors Journal, 16*, 1969–1974.
139. Jung, H.-J., Park, J., & Kim, I.-H. (2012). Investigation of applicability of electromagnetic energy harvesting system to inclined stay cable under wind load. *IEEE Transactions on Magnetics, 48*, 3478–3481.
140. Lu, J.-J., Yang, X.-X., & Mei, H. (2016). A four-band rectifier with adaptive power for electromagnetic energy harvesting. *IEEE Microwave and Wireless Components Letters, 26*, 819–821.
141. Mallick, D., Amann, A., & Roy, S. (2016). High figure of merit nonlinear microelectromagnetic energy harvesters for wideband applications. *Journal of Microelectromechanical Systems, 26*, 273–282.
142. Nintanavongsa, P., Muncuk, U., & Lewis, D. R. (2012). Design optimization and implementation for RF energy harvesting circuits. *IEEE Journal on Emerging and Selected Topics in Circuits and Systems, 2*, 24–33.
143. Ali, M., Albasha, L., & Qaddoumi, N. (2013) RF energy harvesting for autonomous wireless sensor networks. In *2013 8th International Conference on Design & Technology of Integrated Systems in Nanoscale Era (DTIS)* (pp. 78–81). IEEE.
144. Kim, S., Vyas, R., & Bito, J. (2014). Ambient RF energy-harvesting technologies for self-sustainable standalone wireless sensor platforms. *Proceedings of the IEEE, 102*, 1649–1666.
145. Krishnan, S. H., Ezhilarasi, D., & Uma, G. (2013). Pyroelectric-based solar and wind energy harvesting system. *IEEE Transactions on Sustainable Energy, 5*, 73–81.
146. Tan, Y. K., & Panda, S. K. (2010). Optimized wind energy harvesting system using resistance emulator and active rectifier for wireless sensor nodes. *IEEE Transactions on Power Electronics, 26*, 38–50.
147. Zhao, L., Tang, L., & Liang, J. (2016). Synergy of wind energy harvesting and synchronized switch harvesting interface circuit. *IEEE/ASME Transactions on Mechatronics, 22*, 1093–1103.
148. Tan, Y. K., & Panda, S. K. (2011). Self-autonomous wireless sensor nodes with wind energy harvesting for remote sensing of wind-driven wildfire spread. *IEEE Transactions on Instrumentation and Measurement, 60*, 1367–1377.
149. Chen, T., Xia, Y., & Liu, W. (2016). A hybrid flapping-blade wind energy harvester based on vortex shedding effect. *Journal of Microelectromechanical Systems, 25*, 845–847.
150. Habibzadeh, M., Hassanalieragh, M., & Ishikawa, A. (2017). Hybrid solar-wind energy harvesting for embedded applications: Supercapacitor-based system architectures and design tradeoffs. *IEEE Circuits and Systems Magazine, 17*, 29–63.
151. Rezaei-Hosseinabadi, N., Tabesh, A., & Dehghani, R. (2014). An efficient piezoelectric windmill topology for energy harvesting from low-speed air flows. *IEEE Transactions on Industrial Electronics, 62*, 3576–3583.
152. Rezaei-Hosseinabadi, N., Tabesh, A., & Dehghani, R. (2015). A topology and design optimization method for wideband piezoelectric wind energy harvesters. *IEEE Transactions on Industrial Electronics, 63*, 2165–2173.
153. Porcarelli, D., Spenza, D., & Brunelli, D. (2014). Adaptive rectifier driven by power intake predictors for wind energy harvesting sensor networks. *IEEE Journal of Emerging and Selected Topics in Power Electronics, 3*, 471–482.
154. Ando, B., Baglio, S., & L'episcopo, G. (2012). Investigation on mechanically bistable MEMS devices for energy harvesting from vibrations. *Journal of Micro Electromechanical Systems, 21*, 779–790.

155. Bouendeu, E., Greiner, A., & Smith, P. J. (2010). A low-cost electromagnetic generator for vibration energy harvesting. *IEEE Sensors Journal, 11*, 107–113.
156. Chamanian, S., Baghaee, S., & Uluşan, H. (2019). Implementation of energy-neutral operation on vibration energy harvesting WSN. *IEEE Sensors Journal*.
157. Kim, J., Kim, J., & Sim, M. (2013). A single-input four-output (SIFO) AC–DC rectifying system for vibration energy harvesting. *IEEE Transactions on Power Electronics, 29*, 2629–2633.
158. Leicht, J., Amayreh, M., & Moranz, C. (2015). 20.6 Electromagnetic vibration energy harvester interface IC with conduction-angle-controlled maximum-power-point tracking and harvesting efficiencies of up to 90%. In *2015 IEEE International Solid-State Circuits Conference-(ISSCC) Digest of Technical Papers* (pp. 1–3). IEEE.
159. Liu, X., Qiu, J., & Chen, H. (2015). Design and optimization of an electromagnetic vibration energy harvester using dual Halbach arrays. *IEEE Transactions on Magnetics, 51*, 1–4.
160. Nguyen, S. D., & Halvorsen, E. (2011). Nonlinear springs for bandwidth-tolerant vibration energy harvesting. *Journal of Microelectromechanical Systems, 20*, 1225–1227.
161. Qiu, J., Liu, X., & Chen, H. (2015). A low-frequency resonant electromagnetic vibration energy harvester employing the Halbach arrays for intelligent wireless sensor networks. *IEEE Transactions on Magnetics, 51*, 1–4.
162. Sang, Y., Huang, X., & Liu, H. (2012). A vibration-based hybrid energy harvester for wireless sensor systems. *IEEE Transactions on Magnetics, 48*, 4495–4498.
163. Sato, T., Watanabe, K., & Igarashi, H. (2014). Coupled analysis of electromagnetic vibration energy harvester with nonlinear oscillation. *IEEE Transactions on Magnetics, 50*, 313–316.
164. Sun, C., Mu, X., & Siow, L. Y. (2014). A miniaturization strategy for harvesting vibration energy utilizing Helmholtz resonance and vortex shedding effect. *IEEE Electron Device Letters, 35*, 271–273.
165. Sun, Y., Hieu, N. H., & Jeong, C.-J. (2011). An integrated high-performance active rectifier for piezoelectric vibration energy harvesting systems. *IEEE Transactions on Power Electronics, 27*, 623–627.
166. Vijayaraghavan, K., & Rajamani, R. (2009). Ultra-low power control system for maximal energy harvesting from short duration vibrations. *IEEE Transactions on Control Systems Technology, 18*, 252–266.
167. Wang, W., Yang, T., & Chen, X. (2012). Vibration energy harvesting using a piezoelectric circular diaphragm array. *IEEE Transactions on Ultrasonics, Ferroelectrics, and Frequency Control, 59*, 2022–2026.
168. Yan, B., Zhang, C., & Li, L. (2015). Design and fabrication of a high-efficiency magnetostrictive energy harvester for high-impact vibration systems. *IEEE Transactions on Magnetics, 51*, 1–4.
169. Yang, Z., & Zu, J. (2015). Toward harvesting vibration energy from multiple directions by a nonlinear compressive-mode piezoelectric transducer. *IEEE/ASME Transactions on Mechatronics, 21*, 1787–1791.
170. Yuan, X., Changgeng, S., & Yan, G. (2016). Application review of dielectric electroactive polymers (DEAPs) and piezoelectric materials for vibration energy harvesting. In *Journal of Physics: Conference Series*. IOP Publishing (p. 012077).

M. Saravanan received BE degree in electrical and electronics engineering from Coimbatore Institute of Technology, Coimbatore, India in 2006 and ME in VLSI design from Anna University Coimbatore, Coimbatore, India in 2009. He is working as an assistant professor in the Department of Electronics and Communication Engineering at SNS College of Technology, Coimbatore, India. His research interests include VLSI design and semiconductor devices.

J. Ajayan received BTech degree in electronics and communication engineering from Kerala University, Trivandrum, India, in 2009, MTech degree in VLSI design from Karunya University, Coimbatore, India, in 2012, and PhD degree in electronics and communication engineering from Karunya University, Coimbatore, India, in 2017. He is a senior assistant professor in the Department of Electronics and Communication Engineering at SNS College of Technology, Coimbatore, Tamilnadu, India. He has presented papers in many international conferences, and also he is an author of many refereed international journals (Elsevier, Taylor and Francis, Springer, IOP Science).

Sathish R. Jondhale is currently working as an assistant professor in Electronics and Telecommunication Department, Amrutvahini College of Engineering, Sangamner, Maharashtra, India for more than a decade. His research interests are signal processing, target localization and tracking, wireless sensor networks, artificial neural networks and applications, image processing, and embedded system design. He has several publications in reputed journals such as IEEE Sensors Journal, Ad Hoc Networks (Elsevier), Ad Hoc Sensor & Wireless Networks, and Wiley. He is also a member of professional societies such as IEEE and ISTE. He has been a reviewer for peer-reviewed journals such as IEEE Sensors, Signal Processing (Elsevier), IEEE Access, IEEE Signal Processing Letters, and Ad Hoc and Sensor Wireless Networks.

P. Mohankumar received his BE degree in electronics and instrumentation engineering from Anna University, Chennai, India, in 2013 and ME degree in mechatronics engineering from Kongu Engineering College, Erode, India, in 2015. Currently, he is working as an assistant professor in the Department of Mechatronics Engineering at SNS College of Technology, Coimbatore, India. His research interests include VLSI design, MEMS, ergonomics, and sensors and instrumentation.

Energy Efficient ACPI and JEHDO Mechanism for IoT Device Energy Management in Healthcare

A. Sampathkumar, S. Murugan, Ravi Rastogi, Manas Kumar Mishra, S. Malathy, and R. Manikandan

1 Introduction

For the medical world, above all else is the need to change the customary techniques into a computerized structure to energize the advertisement of therapeutic (e-restorative) gadgets, including wearable/versatile/handheld gadgets, which are the need of the hour for data sharing/change. Along these lines, the IoT is received mutually with PLM as a conclusive and savvy approach to business for adequately dealing with the IoT life-chain-based gadgets at the time of treating the patient in the medical emergency clinics/focuses/theaters [1]. At the end of the day, almost every part of human lives, for example, processing plants to ranches, urban areas to towns and medical services to prosperity, and so on, will be cached consistently in 2020, so PLM innovation is no exception. This examination only assigns the utilization of the IoT in human services application, known as the IoMT/MIoT or Health care IoT.

A. Sampathkumar (✉) · M. K. Mishra · R. Manikandan
School of Computing Science and Engineering, VIT Bhopal University, Bhopal, India
e-mail: sampath.kumar@vitbhopal.ac.in

S. Murugan
Department of Computer Science and Engineering, Mewar University, Chittorgarh, India

R. Rastogi
The Faculty of Computing and Information Technology, University of Bisha, Bisha, Saudi Arabia

S. Malathy
Department of ECE, Sri Krishna College of Technology, Coimbatore, India

G. R. Kanagachidambaresan et al. (eds.), *Internet of Things in Smart Technologies for Sustainable Urban Development*, EAI/Springer Innovations in Communication and Computing, https://doi.org/10.1007/978-3-030-34328-6_8

PLM is a viable administration concern for an efficient value chain supply to medical organizations and enterprises. At present, the quickly shifting social insurance situation and extremely intense challenge from partners incites lower development and profit edges, while PLM is trying to oversee boundless wearable gadgets and their networks. Be that as it may, the emergency clinics and medical performance centers are taking a gander at progressively all-encompassing methodologies to bring best-in-class innovations into social insurance businesses to quicken assessment and care giving offices at financially savvy rates. The result of enormous provisos between accessible mechanical ranges of abilities and wanted social insurance requests is exceptionally lumbering.

In addition, it is seen that an enormous number of medical human services organizations are not appropriately dealing with the cross-useful coordinated effort that is urgently expected to amend and bolster customary social insurance advertising conditions. One generous way to deal with the previously mentioned human services difficulties is the joint IoT and PLM-based set ready for overseeing well-being related data over the therapeutic gadgets, emergency clinics and doctors, and so forth [2]. Additionally, our proposed arrangement gives medical emergency clinics, focuses, and associations the capacity to team up and share data rapidly and productively. By thinking about this inclination as the need of the hour, some key answers are found for overseeing the well-being business sector's exhibition as far as lifecycle of battery and power assets for little IoT gadgets. The IoT innovation is as of now changing the standard of the game by creating therapeutic gadgets and fixing affirmation and enhancing the future patterns in medical gadgets with the huge job of PLM innovation in dealing with their availability.

Similar to the expanded intricacy and availability of these versatile gadgets, it is important to embrace the PLM to deal with the life cycle of the battery and its energy from the emergency clinics to the restorative ventures. The usage of PLM is a necessity for minute and tremendous undertakings in view of the sharing capacity of data among various times of the lifecycle. Every piece of mission-essential information is easily managed by PLM, which impacts the working standard of the therapeutic devices. By and large, PLM attempts a methodology of protecting joint exertion and correspondence between restorative pros, business pioneers, designers, experts, and other fundamental requests [3]. Owing to the light weight, accommodating, versatile nature of wearable sensor contraptions in the IoT, the amount of related equipment is set to outperform people. The world plan of the comprehensively related therapeutic contraptions is commonly called the IoMT. However, the confined battery lifetime of wearable sensor contraptions currently could not accomplish the necessities of the restorative globe.

In this way, advancement of the life cycle of battery and energy on the board methods is extremely significant. Additionally, along with the perceiving and requesting job of the IoT innovation, PLM has likewise turned into the focal point of consideration and is at the core of the advanced therapeutic undertakings and ventures.

2 Related Works

2.1 *The IoMT and ACPI for Healthcare Applications Framework*

Innovation of the IoT has distinct advantages and is at the center of consideration for remolding the medical enterprises and conventional companies to a computerized structure. In parallel, PLM is not a long ways behind in totally changing the therapeutic scene, by assuming the job of the driving force in managing, sending, and using restorative items (i.e., wearable equipment). PLM and the IoT are sometimes mutually utilized for tangled systems of programming, correspondence, board parts, etc.

To utilize data returning from the IoT, new potential is needed for PLM; for instance, accessibility as well as scale for orchestrating by means of the amassing engineers, helpful aces, emergency focuses, and patients. For the suitable and skilled treatment of patients through data exchanging and getting the hang of sharing among patients and masters, there is a need for a valid and front line plan [4]. This chapter proposes the novel scheme of a joint IoT and PLM system for successful therapeutic applications. The framework proposed contains four significant advances. In any case, e-Cloud with IoT empowered gadgets is made, in which contraptions can converse with one another by distributing and exchanging essential data.

Then, the essentialness and lifecycle of these contraptions is coordinated from the timeliest starting stage to the arrangement process for treatment by following the stack layer of IoT gadgets. Similar to the way that emergency office information can be dealt with in the IoT-based PLM server, the data will be cleared on the notice sheets of the restorative places for the straight imposition and solace of both the patients and specialists as appeared in the fourth experience of the proposed structure. By following the evacuated information, an ace/master will be apportioned to the patients, time and dates as well as room information can be passed on, and remedy estimations will be doled out to emergency patients.

2.2 *Battery and Energy Management Life Cycle*

IoMT rising patterns and network confronting are two major issues. To begin with, the availability for data sharing among a few gatherings, for example, from assembling engineer to medical master, from the point of clinics/medical services focused on malady analysis of the patient and their treatment. Then, the life cycle of battery and energy of the executives of wearable/compact gadgets are of concern [5].

Here we attempted to take care of the primary issue by coordinating PLM with the IoMT to deal with moving data from one element to another and between gadgets. The next issue is settled by proposing Joint Energy Harvesting and Obligation Cycle streamlining-based (JEHDO) algorithms to improve the lifecycle of battery and energy consumption resources.

2.3 Lifecycle Management in the IoMT System for Battery Recovery-Based Algorithm

The IoMT is the social affair of the wearable, helpful devices/sensors, which actively use batteries. It is hard in many cases to override the battery of these benefit constrained devices, so battery life cycle administration techniques require fixing. Here the presented procedure for battery lifetime in sensor centers is connected by using battery recovery in the IoMT [6]. A battery charge recuperation model is relied upon to recoup the lost charge and administer the lifecycle from beginning to end point. In the control recuperation instrument, closed off charge is circumvented resulting in embedded idle time (δ) between assignments during information/data trading/sharing over joint PLM and the IoMT.

3 Proposed Methodology

3.1 Model for Battery Framework

A systematic battery model is an essential and astounding element for breaking down and understanding the instrument of the minor gadgets while transmitting or imparting data to proceed with cooperation. Broad research has only been directed at battery displaying, highlighting removal and extraction. In addition, diminishing energy utilization is crucial for the battery-fueled wearable sensor gadgets that will be a part of things to come in the IoT [7].

At present, the IoT frameworks are surveyed by thinking about the battery as an energy holder [1]. The two needed components are the present burden and the obligation cycle. In batteries, the burden profile is spoken to as the intermittent arrangement of the N consistent current qualities $I1, I2, I3, \ldots, IN$; though, Ik is the current of assignment k at time tk, and it is utilized for a between entry length time $\Delta k = tk + 1 - tk$, which is the relationship between the current and time $\{Ik, tk\}$ and the battery lifetime L.

The battery achievement will be shown and inaccessible charge will be changed over into accused ones of expanded estimation of β. To discover the battery charge use $\sigma(t)$ after the arrangement of $M(M < N)$ at time $(T < L)$, change N with M and L with T, uncovering the measure of undertakings and T is the due date time for completing undertakings. The battery's utmost cost $\sigma(t)$ merges two fragments, one is gobbled up charge $C(t)$ and the other is inaccessible charge $U(t)$.

$$\alpha(t) = \sum_{k=1}^{N} I_k \Delta_k + \sum_{k=1}^{N} 2I_k \sum_{m=1}^{\infty} \frac{\mathrm{e}^{-\beta^2 m^2 (L - t_k - \Delta_k)} - \mathrm{e}^{-\beta^2 m^2 (L - t_k)}}{\beta^2 m^2}$$

$$\sigma(t) = \sum_{k=1}^{N} I_k \Delta_k + \sum_{k=1}^{N} 2I_k \sum_{m=1}^{\infty} \frac{\mathrm{e}^{-\beta^2 m^2 (T - t_k - \Delta_k)} - \mathrm{e}^{-\beta^2 m^2 (T - t_k)}}{\beta^2 m^2}$$

3.2 The IoMT Algorithm for Energy Harvesting and Duty-Cycle Optimization

One of the rising systems administration models that decrease the float between the genuine and the conjured up universes is the IoMT, in which keen articles speak with one another, information is gathered, and sure requests of the end clients are satisfied as a result. Notwithstanding, a huge segment of the customary arrangements are awesome and expensive from both the computational and the memory resources viewpoint and, as such, they are not often implementable on resource obliged devices. Therefore, vitality gathering methodologies/estimations are an appropriate decision for both vitality and lifecycle of the administrators and, subsequently, the more extended transmission of the sensor center points in medical services condition.

$$\mathrm{DC_S} = \frac{T_{\mathrm{tranON}} + T_{\mathrm{Act}} + T_{\mathrm{tranOFF}}}{T_{\mathrm{tranON}} + T_{\mathrm{Act}} + T_{\mathrm{tranOFF}} + T_{\mathrm{slp}}}$$

where

$$T_{\mathrm{tranON}} = \text{sleep to idle mode Transition time}$$
$$T_{\mathrm{tranOFF}} = \text{idle to sleep mode Transition time}$$
$$T_{\mathrm{Act}} = \text{Active time of a node}$$
$$T_{\mathrm{slp}} = \text{node sleep time}$$

The joint commitment cycle and vitality procuring framework model works in three phases: first to gather vitality from the access point (AP), second to store that vitality in the battery, and third to utilize it to identify and correspond over the IoMT structure.

4 Result and Discussion

Here, the close examination of the proposed ACPI and JEHDO vitality and life cycle of battery in executive calculations with their significant roots in PLM and the IoMT structure are presented. The IoMT shows short division incorporation in which power utilization of intensity speaker (PA) deals with the power channel of handset. The power dissemination of PA (PPA) is implied as transmission control (PTx) and proficiency of PA (ηPA) (Fig. 1; Tables 1 and 2).

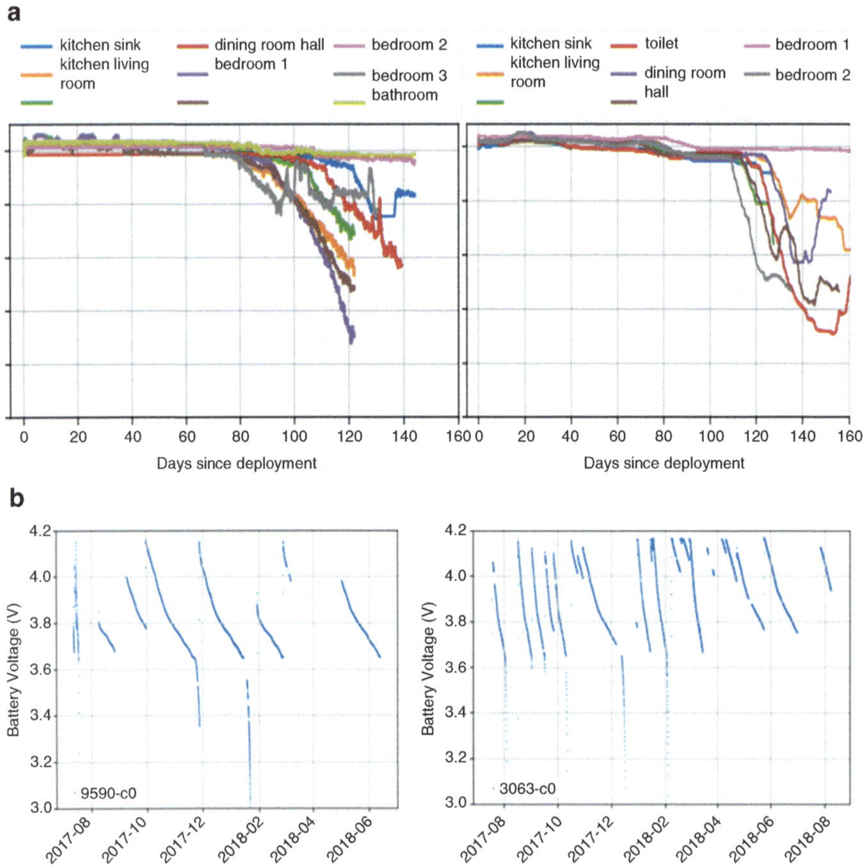

Fig. 1 (**a**) Life time battery estimation using ACPI and JEDHO methods. (**b**) The longest recorded battery lifetime with a single charge for wearable sensors

Table 1 Daily consumption of environmental sensors

Parameter factors	Value
Average energy consumption	8.6 mAh
Node Max. abs. deviation	4.5% (STD = 0.251)
Max. abs. deviation between single node time periods	1.7% (STD = 0.085)

Table 2 Lifetime of environmental sensors

Parameter	Value
Average lifetime, prediction before deployments	780 days
Average lifetime, prediction after deployments	275 days
House 2680 median lifetime	=130 days
House 3619 median lifetime	=150 days

5 Conclusion

This research addresses the first issue of Product Lifecycle Management (PLM) coordinating with the IoMT, which is to direct the movement of data starting with one element then onto the next and between devices in an effective and precise way. The next issue is settled by proposing the battery recuperation-based calculation ACPI and JEHDO calculation to deal with the life cycle of the battery and the energy of the asset compelled small wearable gadgets, individually. Moreover, a novel scheme called joint PLM and IoMT-based structure is introduced for medical healthcare applications. Exploratory results uncover that the Advanced Configuration and Power Management Interface (ACPI) and JEHDO are individually battery-proficient and energy-productive, preparing capacity needs to meet higher and stricter necessities, and the relating equipment expenses of the intensity framework are also tremendous.

References

1. Georgieva, V., & Ermakov, S. (2016). GUI for CT image segmentation via active contours. In *IEEE 4th International Black Sea Conference on Communications and Networking (BlackSeaCom 2016)* (pp. 1–5). https://doi.org/10.1109/BlackSeaCom.2016.7901555.
2. Zhou, W., Bovik, A., Sheikh, C., Hamid, R., & Simoncelli, E. (2004). Image quality assessment: From error visibility to structural similarity. *IEEE Transactions on Image Processing, 13*(4), 600–612.
3. Jayanthiladevi, A., Murugan, S., & Manivel, K. (2018). Text, images, and video analytics for fog computing. In *Handbook of Research on Cloud and Fog Computing Infrastructures for Data Science*.
4. Hegenscheid, K., Kühn, J. P., Völzke, H., Biffar, R., Hosten, N., & Puls, R. (2009). Whole-body magnetic resonance imaging of healthy volunteers: pilot study results from the population-based ship study. *Rapid Communication, 181*, 748–759. https://doi.org/10.1055/s-0028-1109510. Stuttgart/New York: Georg Thieme Verlag KG.

5. Behrad, A., & Masoumi, H. (2010). Automatic spleen segmentation in MRI images using a combined neural network and recursive watershed transform. In *IEEE 10th Symposium on neural network applications in Electrical Engineering, Belgrade, Serbia.*
6. Christensen, G., Rabbitt, R., & Miller, M. (1996). Deformable template model using large deformation kinematics. *IEEE Transactions on Image Processing, 5*(10), 1435–1447.
7. Getreuer, P. (2012). Chan and Vese segmentation. In *Image Processing On Line, IPOL 2012-08-08* (pp. 214–224). Yale University.

A. Sampathkumar received his Bachelor's in Information Technology, Master's in Mainframe Technology, and Ph.D. degree at Anna University Chennai. He is currently working as an Assistant Professor in the School of Computing Science and Engineering at VIT Bhopal University. He has published several articles in peer-reviewed journals and is a member of CSI societies. His research interests include Artificial Intelligence, Data Mining, Machine Learning, Data Analytic, and Optimization Techniques.

S. Murugan obtained his Bachelor's degree in the department of Computer Science and Engineering from Mailam Engineering College, Anna University. He then obtained his post graduate degree in the department of Computer Science and Engineering from Hindustan University, Chennai. Currently, he is Pursuing his Ph.D at Mewar University, Rajasthan. His specializations include Wireless Sensor Networks, Network Security, Cryptography, Database Management Systems and Data Mining. He has published papers in SCi Indexed, Scopus Indexed, other international journals and conferences.

Ravi Rastogi received his Ph.D. degree in Computer Science and Engineering in December 2011 from Uttarakhand Technical University, Dehradun, Uttarakhand, India, the government university. MS in Computer Science degree in September 2005 from Fairleigh Dickinson University, Teaneck, New Jersey, USA. Bachelor of Science degree in 1999 from Ch. Charan Singh University, Meerut, Uttar Pradesh, India. Worked as database developer at 'Microsoft, Seattle, USA' and 'ING, Philadelphia, USA'. Taught and worked at Jaypee University of Information Technology, Waknaghat, Solan, Himachal Pradesh, India for almost 4 years. Taught and worked as department coordinator in 'Sharda University, Greater Noida, Uttar Pradesh, India for 4 years' and 'Galgotias University, Greater Noida, Uttar Pradesh, India for more than 2 years'. Now working at the University of Bisha, Bisha, Saudi Arabia. Research interests are in the fields of Interconnection Networks, Stable Matching Problems, Data Mining and Intelligence in Systems, Internet of Things, and Cloud Computing.

Manas Kumar Mishra completed his B. Tech from Shivaji University, Kolhapur (Maharastra, India), M.Tech and Ph.D (CSE) from Motilal Nehru National Institute of Technology Prayagraj, (Uttar Pradesh, India) in Computer Science & Engineering and Affiliated with Rajiv Gandhi Technical University, Bhopal. He has published many research papers in the area of Wireless Sensor Networks, Mobile Ad-hoc Networks and VANET in international journals and international conferences. He has also published book chapters and edited many Conference Proceedings as the Publication Chair. He has around 21 years of teaching experience in various reputed engineering institutes and universities. He has been involved in organizing several international conferences and workshops. He was the Conference Technical Chair and Publication Chair of several IEEE conferences. He is also the reviewer of many reputed Journals like IEEE Access, Wireless Personal Communications, Springer, KSII Transactions on Internet and Information Systems, Korean Society for Internet Information and Sadhana—Academy Proceedings in Engineering Sciences, Springer.

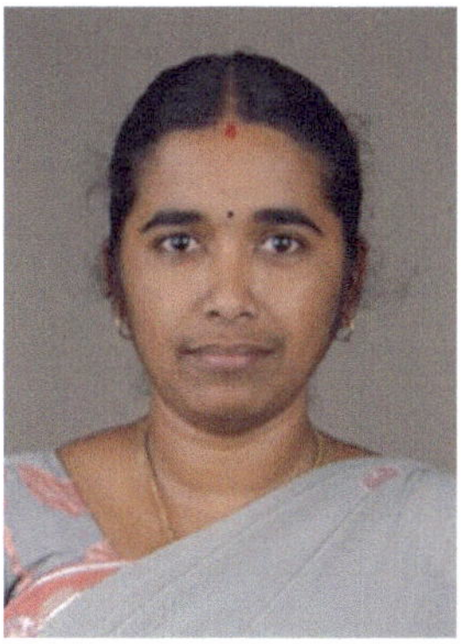

S. Malathy Professor at Sri Krishna College of Technology, Coimbatore, India, received her PhD in Wireless Communication from Anna University in 2013. She has about 20 years of experience in Teaching and Industry. Her areas of interest include Wireless Communication, Wireless Sensor Networks, and Cellular networks. She has published more than 35 papers in various international journals and conferences.

R. Manikandan is working as an Assistant Professor (Senior Grade) in the School of Computing Science and Engineering at VIT Bhopal University, India. He received a B.E. degree in Computer Science and Engineering from the Periyar University, Salem, India, in 2004, and an M.E. in Computer Science and Engineering and PhD degree from Anna University, Chennai, India, in 2007 and 2017, respectively. He is specialized in the area of Cloud Computing, Natural Language Processing, and Data Structures and Algorithms. He is involved in a few funded projects and Consultancies works. He has published various research papers in national and international journals.

Internet of Things (IoT) Based Bio-inspired Artificial Intelligent Technique to Combat Cybercrimes: A Review

M. Balasaraswathi, V. Sivasankaran, N. Akshaya, Radika Baskar, and E. Suganya

1 Introduction

As a part of putative astuteness, two specifics are almost accepted by everyone about the forthcoming century and we may portray them as: the twentieth century was considered to be the century of physics and it is deliberately mentioned that, twenty-first century will be the century of biology. Biology persists as the biggest part of science and is also more important than physics, as measured by its economic sequences, by its ethical insinuation, by its effects on human welfare, by the size of budgets, by the size of workforce, or by the output of major detections, through the twenty-first century [1]. From history, it would be noticed that, for architects and designers, nature has been an inspirational source for various sort of procedures, functions, and techniques. Antediluvian philosophers of Greece were enthused

M. Balasaraswathi
ECE, Saveetha School of Engineering, SIMATS, Chennai, India
e-mail: balasaraswathim.sse@saveetha.com

V. Sivasankaran (✉)
ECE, Sreenivasa Institute of Technology and Management Studies, Chittoor, Andhra Pradesh, India

N. Akshaya
Accenture, Gateway Office Parks, Chennai, India

R. Baskar
ECE, Saveetha School of Engineering, SIMATS, Chennai, India
e-mail: radhikabaskar@saveetha.com

E. Suganya
School of CSE, VIT Bhopal University, Bhopal, India

G. R. Kanagachidambaresan et al. (eds.), *Internet of Things in Smart Technologies for Sustainable Urban Development*, EAI/Springer Innovations in Communication and Computing, https://doi.org/10.1007/978-3-030-34328-6_9

by the organisms, which made them to be inspired with the proportion between their parts. They possess a fascinating concord as seamless models and they had remained a classical ideal of exquisiteness on those days. No trifling part could be detached without detrimental and deform of the whole organism since the quality of integration of the organism's forming parts are tantamount with the edifice, unanimity, and good looks of any design. The way the architects and designers understood nature was so trivial at that juncture. Biology is viewed as a cause of creativeness from the blastoff of science by them and to impersonate and emulate the forms of animals and plants they took an apparent approach. However, nature has been understood as analogies and approaches of development and progression. Understanding of science and nature must be in an unfathomable way but not in just imitating way imaginative depiction, when we talk about analogies. The constraints imposed by the environment, in order to create optimized biological systems in terms of mechanical resistance, mass and life span, is taken into account by nature, for each life style.

For real world optimization problems, it is hard to find the optimized way with multiple objectives. Maximum of these problems are encountered as NP-hard problems and using the traditional deterministic algorithms it could not be solved. The searching process (used to pinpoint top key to a problem which is considered to be optimization) can be done using agents. Based on specified rules of mathematical models, the agent does the searching process iteratively. Ever since the real world problems are mostly escorted with qualms, the optimal solution search is quite complex. As a consequence, more attention is obligatory to have optimum values and burliness of the values that have high impact in engineering design and industries [2]. To address these complex optimization problems, bio-inspired (BI) algorithms have been established to be exceptional methods. BI algorithms can be used to crack various snags that belong to diverse dominions. Numerous BIs have been technologically advanced with the stimulation from the processes that occur in nature such as biological swarms, natural evolution, physical, chemical, and geographic methods over the past few decades [3–5].

BI algorithms turn into a new epoch to compute extensive range of applications in divergent domains such as bio-medical engineering, control systems, parallel processing, data mining, big data, power systems, production engineering, robotics, computer networks, security, mechatronics, and many more fields [6]. No saturation limit (in the sense of presentation along with improvement), flexibility (in the sense of magnitude) are few advantages in bio-inspired algorithms. To amend with easy implementation, BI algorithms possess limited parameters. Since BI algorithms aim to deport themselves well to attain the objective rapidly, it also has certain shortcomings in component design, poor adaptability, and low performance [7]. BI algorithms are cogitated as a major subsection of nature inspired (NI) algorithm [8]. Figure 1 shows the broad classification of NI algorithms.

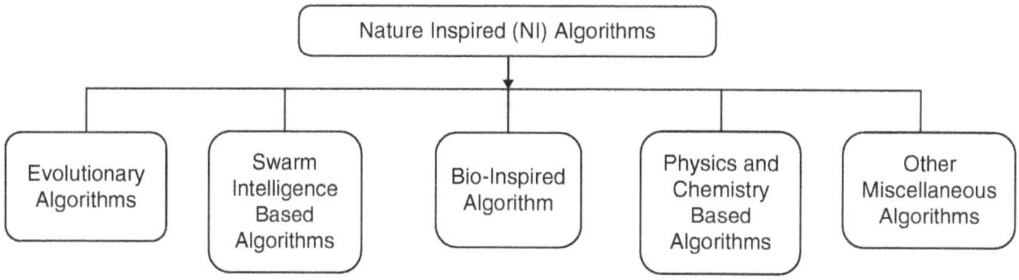

Fig. 1 Broad classification of nature inspired (NI) algorithms

1.1 Evolutionary Algorithms

To rectify difficult, real-time optimization hitches, the evolutionary algorithms are extensively used. Evolutionary algorithms are further classified as: genetic algorithm (GA), genetic programming (GP), evolutionary strategy (ES), and differential evolution (DE). The initial base for the accomplishment of genetic algorithms was formed by Schema theory [9, 10]. Selection, cross-over followed by mutation seem to be significant operators used in GA. With incessant improvement of the solution, DE technique optimizes a given function. GP is a well-known evolutionary technique where in a basic genetic program, dynamic sized tree of different values and functions are employed. A fixed-length, real-valued vector, in which, each vector epitomizes a feature of the individual is used in ES.

1.2 Swarm Intelligence (SI)-Based Algorithms

The wholesome, incipient deportment of numerous, cooperating agents who abide by few meek directions is the one which is concerned by the swarm intelligence. Encouraged by the collective deeds of many social organisms such as ants, birds, fishes, bees, wasps, termites, fireflies, cats, and monkeys, all SI-based algorithms use multi-agents. The swarming activities of fish and birds are used in classical particle swarm optimization (PSO) algorithm, while flashing behavior found in swarming fireflies are used by firefly algorithm (FA). Bat algorithm uses the echolocation of foraging bats while cuckoo search (CS) is based on the brooding parasitism of some cuckoo species. Bee algorithms are based on the foraging behavior of honeybees and ant colony optimization (ACO) uses the interaction of social insects (ants). SI-based algorithms have gained more popularity to unravel a given optimization problem. Many reasons are there behind the fame of SI-based algorithms: first cause is that SI-based algorithms distribute data to numerous driving forces—such that self-structuring, concurrent-progress and studying, during the iterations, might pave way to give good performance; second cause is that the poly agents could be made parallel so that from the execution view point, optimization in a large-scale manner would be suiting well to the real time [11, 12].

SI-based algorithms can be enumerated (numbered) as follows: Ant Colony Optimization (ACO) [13], Accelerated PSO [14], Artificial Bee Colony [15], Bacterial Foraging [16], Bacterial GA Foraging [17], Bat Algorithm [18], Bee Colony Optimization [19], Bee System [20], Bee Hive [21], Wolf Search [22], Bees Algorithms [23], Bees Swarm Optimization [24], Bumblebees [25], Cat Swarm [26], Consultant-guided Search [27], Cuckoo Search [28], Eagle Strategy [29], Fast Bacterial Swarming Algorithms [30], Firefly Algorithm [31], Fish Swarm/School Algorithm [32], Good Lattice Swarm Optimization [33], Glowworm Swarm Optimization [34], Hierarchical Swarm Model [35], Krill Herd [36], Monkey Search [37], Particle Swarm Algorithm [38], Virtual Ant Algorithm [39], Virtual Bees [40].

1.3 Bio-inspired (But Not SI-Based) Algorithms

Numerous algorithms that are inspired by biological phenomenon other than swarm behavior exist, where SI-based algorithms is a narrowed down view of a larger class, named bio-inspired algorithms. However, main stream of all NI algorithms are formed by BI algorithms. If viewed from the point of set theory, SI-based algorithms are a subgroup of BI algorithms and BI algorithms are a subcategory of NI algorithms. Inversely speaking, not all the NI algorithms are bio-inspired and some are virtuously physics- and chemistry-based algorithms. The swarming behavior is not directly used by many BI algorithms, and hence it may be termed as bio-inspired, but not SI-based algorithms. For an example, genetic algorithms are BI, but not SI-based and likewise, DE is not BI since no straight link is established to any biological behavior. As DE has some similitude to genetic algorithms and also has a key word "evolution," it has been nearly kept in the category of BI algorithms. Pertinent algorithms are: Atmosphere Clouds Model, Bio-geography Based Optimization (BBO), Brain Storm Optimization (BSO), DE, Dolphin Echolocation, Japanese Tree Frogs Calling, Eco-Inspired Evolutionary Algorithm, Egyptian Vulture, Fish-School Search, Flower Pollination Algorithm (FPA), Gene Expression, Great Solomon Run, Group Search Optimizer, Human-Inspired Algorithm, Invasive Weed Optimization, Marriage in Honey Bees, Opt Bees, Paddy Field Algorithm (PFA), Roach Infestation Algorithm, Queen-Bee Evolution, Shuffled Frog Leaping Algorithm (SFLA), and Termite Colony Optimization (TCO) [12].

1.4 Physics- and Chemistry-Based Algorithms

A category of set of rules that are generally motivated by the physical or chemical processes befalling under a description, named as "physics and chemistry based algorithms." Since all fundamental laws are same, no need to classify again into physics-based and chemistry-based algorithms. By replicating some physical and/or chemical laws (which includes electrical charges, gravity, river systems, etc.), these algorithms have been created and a variety of algorithms that belongs to this category are: Big Bang-Big Crunch, Black Hole, Central Force Optimization, Charged System Search, Electro-Magnetism Optimization, Galaxy-Based Search Algorithm, Gravitational Search, Harmony Search, Intelligent Water Drop, River Formation Dynamics, Self-Propelled Particles, Simulated Annealing, Stochastic Diffusion Search, Spiral Optimization, Water Cycle Algorithm [12].

1.5 Other Miscellaneous Algorithms

While new algorithms are developed, some algorithms may fall under BI category, some may be physics- or chemistry-based and others may seek encouragement further than nature; hence it is often challenging to place some algorithms in the above mentioned three divisions, because these algorithms have been established by employing various behaviors from numerous bases, like social, emotive, and so on. Therefore, it might be good to categorize them under "other" domain. Few algorithms that represent this category are: Anarchic Society Optimization, Artificial Cooperative Search, Backtracking Optimization Search, Differential Search Algorithm, Grammatical Evolution, Imperialist Competitive Algorithm, League Championship Algorithm and Social Emotional Optimization [12].

The bases of encouragement, which are very different, from where the algorithms are conceived may also be similarly different. For instance, if we consider that there are about "*N*" number of living species of fish, it is said not with the intention that the researchers would design "*N*" number of different algorithms that are based on "fish" and no one can term their algorithms as squid algorithm, trout algorithm, shark algorithm, etc. Some algorithms may lack in some fundamental competences like diversity and mixing among the solutions. But some algorithms may possess these abilities and efficiently explore the search space and converges rapidly at the required time. Algorithms that have wide range of applicability, global search, and intensive local search abilities are considered as efficient and good; among them PSO, DE, cuckoo search, and firefly algorithms are found to be very efficient algorithms. The notion behind swarm intelligence is that, the algorithms that are developed need to be malleable for both internal and external changes, to be robust when few entities fail, to be distributed and self-organized and this inkling involves in working on artificial intelligence.

2 Artificial Intelligence (AI) and Bio-inspired Algorithms

A ground of processing which manages the subfields accompanied with the issues of connectionism, emergence, and social behavior are, as we know, bio-computing or biologically inspired computing. Since many of BI algorithm's mission is able to be associated to machine learning, it is narrowly interrelated to the branch of artificial intelligence and also it prominently bank on the fields of mathematics, computer science, and biology. A twofold delineation can be applied to the idea of BI algorithms: utility of ecology or natural processes as allegory, motivation or designer to develop new areas of computer science and innovative computing technologies and to make use of the concepts of information science and mechanisms to leave no stone unturned in biology with a contradistinctive metaphysical context. BI computing has connotations for rudimentary (basic) scientific research along with its prospective uses such as nanofabrication, DNA computation, storage devices, sensing, and health care. For precedent, BI algorithms can help computer scientists to construct algorithms based on NI approaches (such as evolutionary and genetic algorithms) and afford biologists with an InfoTech-oriented exemplar to look at the way of computation and processing of information by cells.

The progress of many sophisticated algorithms have aggravated from the ideas of natural and biological deeds. As mentioned earlier, two significant divisions of NI algorithms are EA and Swarm Intelligence (SI) algorithms. Survival of the fittest or natural selection principles provides inspiration to EA to particularize algorithms relevant to that, whereas, the amassed intellect of swarm or behavior of insect and other animal societies contribute stimulation to SI algorithms for algorithms and distributed problem solvers. SI can be demarcated as relatively new division of AI because they have the ability of giving fast, robust, cost-cutting findings to harder real-time issues. Bio-inspired algorithms take more evolutionary approach to learn and in conventional AI, intelligence is programmed from BI algorithms. The programmer, who is the creator of program, infuses AI with its intelligence in some way. Mentioning a set of easier procedures, a set of simple organisms that obey to those procedures and repetitively deploying the procedures are the methods involved in BI techniques. But, the science of imparting intelligence in machines which in turn performs the tasks (where human mind is required) is baptized as artificial intelligence. In two ways, AI can be described: one as a science that targets to discover the quintessence of intelligence and develop intelligent machines and another one as a science that finds strategies to solve complex problems by applying some intelligence. The second way is more suitable when AI is applied for the defense of cybercrimes.

2.1 History of Cybercrime and AI

Mankind has been motivated ever since the beginning of civilization to make tremendous development and better the existing technologies and most significant among them is Internet. It is believed that internet is full of mayhem which makes a system of law and regulation to be conflicting. A system of law called "cyber law" is being governed by "cyberspace." All legal and regulatory aspects of internet is referred by a generic term "cyber law," which evolves constantly. Growth of internet creates numerous legal issues and the most important problem among cyberspace is cybercrime. As per Halder and Jaishankar (2011), using modern telecommunication networks such as internet (chat rooms, emails, notice boards, and groups) and mobile phones (SMS/MMS), when offences are caused counter to a single person or collection of persons with a criminal mentality for the sake of intentionally hurting the respect of the victim or to create physical or mental danger to the sufferer openly or in a roundabout way, is termed as "cybercrime."

In the year 1820, the first recorded cybercrime took place. Though abacus, which is considered as the most primitive form of a computer, was in practice in India, China, and Japan since 3500 BC, the epoch of modern computers originated with Charles Babbage's analytical engine. A textile manufacturer, named Joseph Marie Jacquard, France, manufactured the loom. The nitty-gritty of this device is it allows the recurrence of sequence of phases while weaving special fabrics. As a result, this created a wave of trepidation, among the employees of Jacquard, that their long-established service and livelihood were being endangered. To dishearten Jacquard from the further practice of new technology, the wage-earners committed acts of disrupt. This is considered as the recorded cybercrime in broader aspect. Nowadays, every atom of water in a glass could be converted into a computer by performing billions of operations per second with neural networks and nano-computing. The progressing necessity of computers in digital life is the origin for the evil of cybercrime.

The distinctive trait of storing information in a very narrow space of a computer, unauthorized, easy and secret access, complex coding which helps cyber criminals to penetrate into the computer system, and loss of evidence are said to be the vulnerabilities of computers. With internet adoption, rate of cybercrimes increases and new levels of vulnerability thus introduced because of mobile internet access and continuous deployment of broadband internet structure. Attempts could be made to elude cybercrimes by identifying phishing, recognizing that the smartphones used by us also remains as a pocketsize computer which is prone to same sort of bouts done to a lap/desktop, keeping the personal information with ourselves and not sharing in social medias, being aware of public Wi-Fi and computers, evading doubtful emails, and checking of accounts and credit reports in regular manner. These verdicts are just illustrations of misapplication of the internet and hence provide the reason for slowing down the cybercrime.

A technology that targets to make a computer to do the human like reasoning is artificial intelligence (AI). Using AI, all human beings, plants, animals, or whatever it may be could be connected together and thus make the world an autonomous place by making "smart decisions." Reasoning, representation, and learning are the three main components of AI. Putting into nut shell, it can be said that, the knowledge of introducing intelligence in machines and making them to perform jobs which are needed for man mind is AI. In the sense of application, adaptation, processing speed and capabilities, the AI systems are developing quickly and availing less-routine tasks. The key difference between human intelligence and AI is that the former "takes" a perfect decision and the latter "chooses" a right decision at appropriate time. Though the creativity in decision that human has may lack in AI, it is often showed that AI lessens replication of human efforts and produces better results.

AI like machine learning, permits the computers to allow for observing, analyzing, self-training, and learning by themselves. However, numerous technical systems seam strengths to work together in a coordinated way, in swarm intelligence. New technologies and strategies such as machine learning and automation are adopted by companies and the cybercriminals may try to use automated fuzzing with machine learning. A technique of attack that applies swarm intelligence, named, "swarm-as-a-service" is anticipated to be used unduly in future. AI techniques have numerous applications in detection and prevention of cybercrime and for example, it is probable to develop highly efficient intrusion detection and prevention using neural networks and new generation antivirus systems use AI to improve their efficiency.

3 IOT and Cyber-Physical Systems (CPS)

An important technology which assures a smart individual's life, by means of permitting communications between objects, machines, and everything along with peoples, is Internet of Things (IOT). The increased usage of internet and mobile devices leads to the risk of cybercrimes. From the internet, the IOT enabled cyber-physical systems (CPS), a joining point between the cyber and physical worlds, is facing threats and vulnerabilities. A virus, worm, Trojan horse, disrupting and denying computer services, software bomb and damage are the reasons for system interference. IOT enabled CPS can be made endangered by broad ways of cyber-attacks from criminals, hacktivists and terrorists, so that cybersecurity has become a critical challenge in IOT. Explosion of IOT devices in all environments, increased utility of wireless connectivity and interaction of all components with mobile and cloud platforms are the key drivers behind this trend. To work for the exact destiny of automation and adaptation, a smarter IOT system will have AI, where voice assistants, robots are few examples of AI enabled IOT.

3.1 Combined Application of BIC and AI in Cybersecurity

The greatest advantage of swarm intelligence is its joint strength which could be well applied in cybersecurity. Like an ant leaves its traces in the way in which it finds the food, the module (Bot) leaves its memory of an attack which is faced by it along the network path.

When other bot is taking over the same path, the knowledge will be injected to it. Same is the case for all other bots. As a final step, the bots will collectively or jointly fight against the cyber threats. Also multiple threats could be identified at the same time by multiple bots individually, and the retaliation would be a collective action. This joint strength is the success of combination of artificial intelligence and swarm intelligence in the field of cybersecurity.

Let us consider the cybersecurity framework designed by Samuel Foni and Luca Ronchini as a part of their project SESAR2020 for a complex and critical environment like Airport. The Airport system consists of many interactive modules which adapt to the changes in environment and with each other with respect to time. It is highly sensitive to attack as it is less possible to protect this complex system's flaws that might be due to human activities or due to IT cyber degradation events.

3.2 *Detriments of Conventional Cybersecurity Strategy to Complicated Systems*

Something more than understanding each and every part of the system is needed, in order to get rid of the attacks. This means that, a wholesome approach is required rather than concentrating on a single impact out of multiple attacks at multiple modules of the system. And the response or decision against the attack should be as quick as possible. These qualities are lacking in the traditional cybersecurity approach which paves the way for a new technology to get engaged.

SESAR2020 project is aiming to incorporate the traditional method along with some additional aspects, which would highly secure the cyberspace. As usual, the traditional modules are going to identify the threat, investigate, respond to it, and actively defend the complex system from threat. But other additional modules planned in a such a way that the criteria to identify, investigate, respond, and react to the threats need to be dynamically changing with respect to time in the event of overcoming the increasing cyber-attacks.

Three main logical blocks are needed to update the criteria dynamically with periodic investigations. They are:

- Cybersecurity audit
- Risk valuation
- Cyber resilience review

3.2.1 Cybersecurity Audit

Having a wholesome idea of the system is enough to perform the audit. It is not necessary to know about the subsystems or any other modules of the system because all these details could be obtained by the following way. Various questionnaires are put forth relating to the submodules of the complex system, various security techniques being currently used, all the points where accesses could be done. With these details, an elaborate analysis must be made to identify the significant assets of the system and access limiting devices for the system. These become the input for next blocks and also a sole segment of the construction.

3.2.2 Risk Valuation

With significant assets of the system on one hand, it is necessary to identify the risks associated with the assets on the other hand. Analysis must be made to determine the probability of threats and vulnerable areas of attack. From this analysis, our first requirement is to increase the security measures so that in future the scope of attack should be reduced. In the same way, the framework should also be planned.

3.2.3 Cyber Resilience Review

This deals with the examination of the make-up details of the various protection systems, level of opposition toward attack and the speediness of the whole framework. It also involves physical testing of the detection mechanisms, notifying devices and defense devices. Further moving on, the verification attack recovery plan, efficiency and effectiveness of the framework. Above all, strong control of the response systems must be made. Further improvement in the framework involves new retaliation measures, new working mechanism, and also new response techniques.

3.3 Incorporating Risk-Based Approach into Conventional Mechanism

Artificial intelligence is the technology which is having the capacity to identify the risk, analyzing it, and updating its knowledge for future predictions. This is very much useful for us to both counteract to the current situation and be ready for further future attacks also. That is why the necessity of artificial intelligence's presence in hand with cybersecurity is increasing. In short, the cyber resilience review activity is the significant tool to enhance the cyber threats lessening process. Swarm intelligence into cybersecurity is yet another step that indeed improves the speed of communication between the operator and the system. It follows a hierarchical approach so that the person will have a singer point of contact and his/her counteraction will influence more than one point of the system. In this hierarchical framework, agents at multiple levels interpret the situation, adapt themselves to current threat, and deploy their semi-rationalization and self-learning capability on a logical basis.

So, entirely getting rid of cyber-attacks is less possible. At most, the artificial intelligence and correlation techniques could be well used so that the threats from any means at any time could be speedily and effectively conveyed to the operator. This enables the operator to intervene at correct time and protect the system.

Top layer	Cutting edge data scrutiny and association, centered entirely on AI E.g.: High-level SIEM (Collects alarm and briefly interprets to the operator)	Queen ant
Intermediate layer	Rub the data assembled by workforce, by means of knowledgeable data scrutiny and association E.g.: Mid-level SIEM, IPS, or EDR systems (alarm generation)	Male and soldier ant
Lowest layer	Processing and collecting massive data E.g.: Log analyzers, IDS systems, cyber probes, automatic scripts	Worker ant
	Environment	

EDR endpoint detection and response, *IDS* intrusion detection system, *IPS* intrusion prevention system, *SIEM* security information and event management

4 Conclusion

Putting all the mechanisms inspired from the biological species in a nut shell will furnish a hand full of approaches to the scientists and researchers in the event of optimizing the traditional, conventional, and well-established way of removing the puzzle knots of complex problems. Computer intelligence is evolving out of AI which in turn mandates the existence of computer as an essential part to make things amicable to the human beings. Since IOT-CPS is a complex and diverse web, it will be difficult to keep track of all unethical incidents. A lot inconvenience can also be caused by even power failure. To monitor its whereabouts at each instant, we may require another AI scheme lying on such AI enabled IOT. Someday, our lives will go on with the control of technology, as we will depend on them for everything, which leads to get democracy of such systems for the populaces. Without getting enslaved by the technology and to control this smart revolution, human beings should have reign over all the man-made smartness, irrespective of the situations and cases.

References

1. Dyson, F. (2007). Our biotech future. *The New York Review of Books*.
2. Haupt, R. L., & Haupt, S. E. (2004). *Particle genetic algorithms* (2nd ed.). New York. ISBN: 0-471-45565-2: Wiley.
3. Gandomi, A. H., & Alavi, A. H. (2012). Krill Herd: A new bio-inspired optimization algorithm. *Journal of Communications in Nonlinear Science and Numerical Simulation, 17*(12), 4831–4845.
4. Fister, I., Fister, I., Jr., Brest, J., & Žumer, V. (2012). Memetic artificial bee colony algorithm for large-scale global optimization. *IEEE Congress on Evolutionary Computation*, 1–8.
5. Yang, X. S. (2009). Firefly algorithms for multimode optimization. *Journal of Stochastic Algorithms: Foundations and Applications, Springer*, 169–178.

6. Binitha, S., & Siva Sathya, S. (2012, May). A survey of bio-inspired optimization algorithms. *International Journal of Soft Computing and Engineering, 2*(2), 137–151. ISSN: 2231-2307.
7. Mahale, R. A., & Chavan, S. D. (2012, December). A survey: Evolutionary and swarm based bio-inspired optimization algorithm. *International Journal of Scientific and Research Publications, 2*(12), 1–7.
8. Gandomi, A. H., & Alayi, A. H. (2012). Krill herd: A new bio-inspired optimization algorithm. *Communications in Nonlinear Sciences and Numerical Simulation, 17*(12), 4831–4845.
9. Goldberg, D. E. (1989). *Genetic algorithms in search, optimization, and machine learning*. Reading: Addison-Wesley.
10. Mishra, A., & Shukla, A. (2016). A new insight into the schema survival for genetic algorithms consisting of chromosomes having bits with different vulnerability for mutation. *International Journal of Control Theory and Applications, 9*(42), 39–44. ISSN: 0974-5572.
11. Kar, A. K. (2016). Bio inspired computing—A review of algorithms and scope of applications. *Journal of Expert Systems with Applications, 59*, 20–32.
12. Fister, I., Jr., et al. (2013, July). A brief review of nature-inspired algorithms for optimization. *Article in Elektrotehniski Vestnik/Electrotechnical Review, 80*(3), 1–7.
13. Dorigo, M. (1992). *Optimization, learning and natural algorithms*. Ph.D. Thesis, Politecnico di Milano, Italy.
14. Yang, X.-S. (2010). *Nature-inspired metaheuristic algorithms* (2nd ed.). Bristol: Luniver Press.
15. Dervis, K., et al. (2007). A powerful and efficient algorithm for numerical function optimization: Artificial bee colony (ABC) algorithm. *Journal of Global Optimization, 39*(3), 459–471.
16. Passino, K. M. (2002). Biomimicry of bacterial foraging for distributed optimization and control. *IEEE Control Systems, 22*(3), 52–67.
17. Chen, T.-C., et al. (2007). A novel optimization approach: Bacterial-GA foraging. In *Second International Conference Innovative Computing, Information and Control—ICICIC'07, IEEE* (pp. 391–391).
18. Yang, X. S. (2010). A new metaheuristic bat-inspired algorithm. In *Nature Inspired Cooperative Strategies for Optimization (NICSO 2010)* (pp. 65–74).
19. Teodorović, D., & Dell'Orco, M. (2005). Bee colony optimization—a cooperative learning approach to complex transportation problems. In *Advanced OR and AI Methods in Transportation: Proceedings of 16th Mini-EURO Conference and 10th Meeting of EWGT (13–16 September 2005)* (pp. 51–60). Poznan: Publishing House of the Polish Operational and System Research.
20. Lucic, P., & Teodorovic, D. (2001). Bee system: Modeling combinatorial optimization transportation engineering problems by swarm intelligence. In *Preprints of the TRISTAN IV-Triennial Symposium on Transportation Analysis* (pp. 441–445).
21. Wedde, H. F., et al. (2004). Beehive: An efficient fault-tolerant routing algorithm inspired by honey bee behavior. In *Lecture Notes in Computer Science, Book Series Vol. 3172 LNCS* (pp. 83–94).
22. Tang, R., et al. (2012). Wolf search algorithm with ephemeral memory. In *7th International Conference on Digital Information Management (ICDIM)* (pp. 165–172).
23. Pham, D. T., et al. (2006). The bees algorithm—A novel tool for complex optimization problems. In *Proceedings of the 2nd Virtual International Conference on Intelligent Production Machines and Systems (IPROMS 2006)* (pp. 454–459).
24. Drias, H., Sadeg, S., & Yahi, S. (2005). Cooperative bees swarm for solving the maximum weighted satisfiability problem. *Journal of Computational Intelligence and Bio inspired Systems, Springer, LNCS, 3512*, 318–325.
25. Padró, F. P. C. (2009). *Bumblebees: A multi-agent combinatorial optimization algorithm inspired by social insect behavior*. GEC-2009 (pp. 811–814).
26. Chu, S. A., et al. (2006). Cat swarm optimization. In *Lecture Notes in Computer Science (including subseries Lecture Notes in Artificial Intelligence and Lecture Notes in Bioinformatics), 4099 LNAI* (pp. 854–858).

27. Iordache, S. (2010). Consultant-guided search: A new metaheuristic for combinatorial optimization problems. In *Proceedings of the 12th Annual Conference on Genetic and Evolutionary Computation* (pp. 225–232).
28. Yang, X.-S., & Deb, S. (2009). Cuckoo search via Lévy flights. In *World Congress on Nature & Biologically Inspired Computing, NaBIC 2009, IEEE* (pp. 210–214).
29. Yang, X.-S., & Deb, S. (2010). Eagle strategy using Lévy walk and firefly algorithms for stochastic optimization. In *Nature Inspired Cooperative Strategies for Optimization (NICSO2010)* (pp. 101–111). Berlin: Springer.
30. Chu, Y., et al. (2008). A fast bacterial swarming algorithm for high-dimensional function optimization. In *IEEE World Congress on Computational Intelligence* (pp. 3135–3140).
31. Yang, X.-S. (2010). Firefly algorithm, stochastic test functions and design optimization. *International Journal of Bio-Inspired Computation, 2*(2), 78–84.
32. Li, X. L., et al. (2002). Optimizing method based on autonomous animate: Fish-swarm algorithm. *Xitong Gongcheng Lilun yu Shijian/System Engineering Theory and Practice, 22*(11), 32.
33. Su, S., et al. (2007). Good lattice swarm algorithm for constrained engineering design optimization. In *International Conference on Wireless Communications, Networking and Mobile Computing, IEEE* (pp. 6421–6424).
34. Krishnanand, K. N., & Ghose, D. (2009). Glowworm swarm optimization: A new method for optimizing multi-modal functions. *International Journal of Computational Intelligence Studies, 1*(1), 93–119.
35. Chen, H., et al. (2010). Hierarchical swarm model: A new approach to optimization. *Discrete Dynamics in Nature and Society.*
36. Mucherino, A., & Seref, O. (2007). Monkey search: A novel metaheuristic search for global optimization. *Data Mining, Systems Analysis and Optimization in Biomedicine, 953*, 162–173.
37. Kennedy, J., & Eberhart, R. (1995). Particle swarm optimization. *IEEE International Conference on Neural Networks, 4*, 1942–1948.
38. Yang, X.-S., et al. (2006). Application of virtual ant algorithms in the optimization of CFRP shear strengthened pre-cracked structures. In *Computational Science–ICCS* (pp. 834–837). Springer.
39. Yang, X. S. (2005). *Engineering optimizations via nature-inspired virtual bee algorithms* (Vol. 3562, pp. 317–323).
40. Ting, T. O., et al. (2012). Weightless swarm algorithm (WSA) for dynamic optimization problems. In *Network and Parallel Computing* (pp. 508–515). Springer.

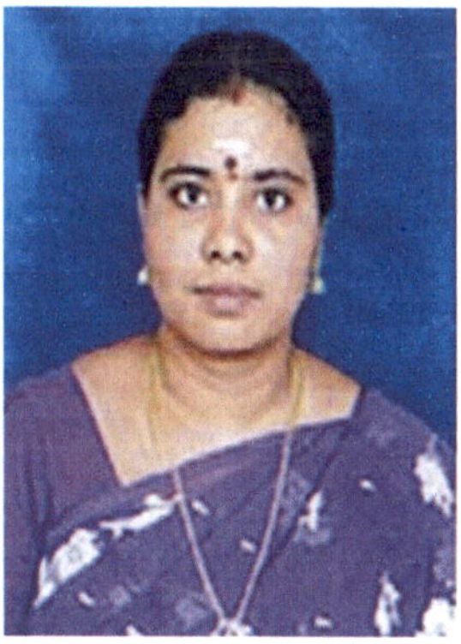

M. Balasaraswathi obtained her bachelor of engineering degree in electronics and communication engineering from University of Madras, in 1999, and a master's degree in applied electronics from College of Engineering, Guindy, Anna University, Chennai, in 2005. She has completed her PhD in information and communication engineering, from Anna University, Chennai, in 2017, in the area of wireless communications. She is presently working as an associate professor in the Department of Electronics and Communication Engineering, Saveetha School of Engineering, SIMATS, Saveetha University, Chennai, India. Her research interests are in the areas of communication systems, signal processing, wireless and sensor networks, and cooperative communication for wireless networks.

V. Sivasankaran obtained his bachelor of engineering degree in electronics and communication engineering from Anna University, in 2005, and a master's degree in applied electronics from Anna University, Chennai, in 2010. He got his doctorate in faculty of information and communication engineering, from Anna University, Chennai, India, in the year 2018. He has more than 13 years of teaching experience. He is presently working as an associate professor in the Department of Electronics and Communication Engineering in Sreenivasa Institute of Technology and Management Studies (Autonomous), Chittoor, Andhra Pradesh, India. His research interest includes communication systems, signal processing, wireless and sensor networks, and HWN.

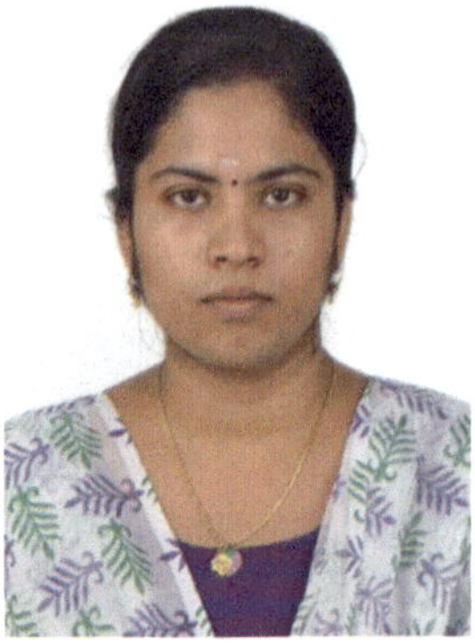

N. Akshaya received her engineering degree at Anna University, Chennai, India, and presently she is working as a software engineer in Accenture, India. Her research interest includes, IOT, wireless communication, and WSN.

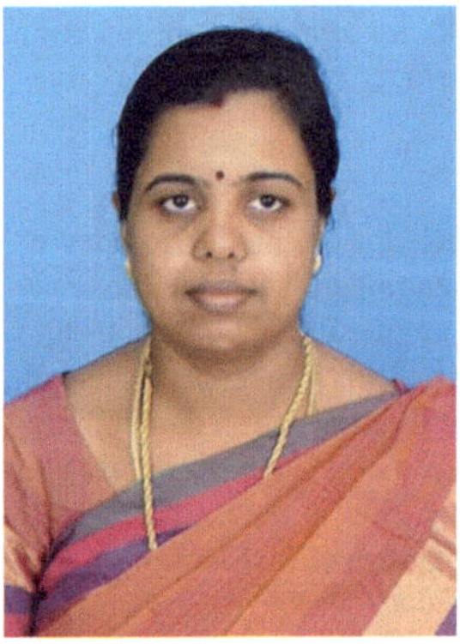

Radika Baskar is working as an associate professor in the Department of Electronics and Communication Engineering at Saveetha School of Engineering, SIMATS. She has around 15 years of teaching experience. She graduated from Vellore Engineering college, Vellore (presently known as Vellore institute of technology, Vellore), post-graduation from College of Engineering, Guindy, Anna University, PhD from Saveetha Institute of Medical and Technical Sciences (SIMATS). She has published 24 papers in referred journals and participated and presented papers in various national and international conferences. Her field of interest includes network security, wireless sensor network, Internet of things, and soft computing.

E. Suganya is presently working as an assistant professor/SCSE at Vellore Institute of Technology Bhopal University, Bhopal, Madhya Pradesh, India. She has done BE-CSE degree from Anna University, Chennai, Tamilnadu and ME-CSE degree from Anna University, Chennai, Tamilnadu, India. Her areas of interest are data mining and machine learning. She has published and presented 12 papers in international conferences and journals and has teaching experience of about 3 years.

Impulsive Parking Systems for Smart Cities Based on Vehicle Location and Destination Using Internet of Things

Mathan Chandran, Susan Immanuel, Thanarajan Kumaresan, R. Suresh Kumar, and K. Vishnu Murthy

1 Introduction

People driving in the busy citified regions would be deeply annoyed to find an empty parking slot and crave for a solution that could lead them to an unfilled parking slot. The hustled movement of the vehicles in search of empty slot frustrates the driver and increases the fuel consumption which in turn raises the pollutant level in the exhaust gas. On searching for an empty slot in bustling urbs, the probability of a mishap is higher [1]. This issue made industries to invest solely on developing smart parking strategies by implementing modern technology. Cities with great ordinal of on-street parking slots comparatively needs to take up smart parking to avoid vehicles rushing to the empty parking slot. Today, parking sector is being changed by modern technologies which facilitate cities to reduce overcrowding of the vehicles and pollutants emission. Internet of Things (IoT) connects the world of parking and provides intelligent parking solutions. In this frame of reference, IoT employs integrated wireless sensor networks to link the parking slots cloud-predicated astute management accommodations are offered. These affiliated devices are termed as smart devices or smart objects, in addition it constitutes smart machines that communicate and interact with each other, environment, objects, etc. Some essential benefits of IoT are (1) easy to track, (2) increased awareness

M. Chandran (✉) · T. Kumaresan · R. Suresh Kumar
Department of Automobile Engineering, PSG College of Technology, Coimbatore, India

S. Immanuel
Department of Chemistry, PSG College of Technology, Coimbatore, India

K. Vishnu Murthy
Department of Electrical and Electronics Engineering, Sri Krishna College of Technology, Coimbatore, India
e-mail: k.vishnumurthy@skct.edu.in

G. R. Kanagachidambaresan et al. (eds.), *Internet of Things in Smart Technologies for Sustainable Urban Development*, EAI/Springer Innovations in Communication and Computing, https://doi.org/10.1007/978-3-030-34328-6_10

about the situation, (3) sensor determined judgmental analytics, (4) rapid control and response, etc. IoT technology grows in diverse fields and the boundary constraints are yet to be identified. To achieve our goal of smart parking, three major procedures are involved namely sensing/detection, processing, and connecting.

The sensing/detection includes the collection of data such as parking information, vehicle order information, and vehicle information by the sensors which forms the heart of the system. Some integrated systems such as Infineon, Freescale, Armcortex, raspberry pi, auridino, Tsmote, Tsgate, etc. are employed as controllers. The data from the sensors can be refined by employing processors such as network processor, hybrid processor, etc. and the contrivances are interlinked by technologies called Wi-Fi, MCU/MPU, GPS, RFID, BT/BTLE, etc. [2]. Smart parking is a vehicle parking system, which is based on road sensors and nimble displays that directs motorists to find a parking space in a circumscribed time. Smart parking system also calculates the time spent by the vehicle in the parking lot. In addition to allocation of unfilled space location and calculate parking payment, the system can also manage parking authorization. Presently, the parking authorizations are printed receipts and digitizing them would be a value add-on which makes the system utilization more feasible.

The contributions of smart parking system include: (1) increasing space utilization, (2) more expeditious time to locate a vacuous parking slot and thereby less CO emission, (3) improving driver's experience, and (4) providing intelligent management. Moreover, by using short-range communication technologies, like near field communication (NFC) or radio frequency identifiers (RFID), it is viable to realize an electronic verification system of parking authorizations in slots reserved for denizens or incapacitated, thus offering a better accommodation to denizens that can legitimately utilize those slots and an efficient implement to expeditiously spot infringements.

From the point of management's view, smart parking is a nimble parking system. The parking method is sculptured as birth-death model and also the foresight of revenues is created. Predicated on the foresight, incipient business promotions can be made. For example, promotion prices (on-sale prices) and newborn parking plans can be publicized and broadcasted to all the passing vehicles without unnecessary costs. In smart parking, new promotions are revealed through wireless network.

This chapter will provide an overview of the various smart parking systems proposed so far. The different sensors and processors used in the system will be elaborated along with a few insights on the future directions toward a successful smart city.

2 Rationale Behind IoT-Cloud Integration

Both Internet of Things and cloud computing have seen a humongous growth in the recent times. Each technology has its own advantages and there are some mutual advantages when integrated. The technological constraints like storage, processing, and efficiency of IoT can be enhanced with the capabilities and resource of cloud computing. By the help of IoT, cloud computing can be used in authentic world entities and in a more dynamic way. Cloud is the link between the things to be connected and the applications. Here are some of the factors for integrating IoT and cloud.

Storage Capacity Internet of Things as the name suggests has numerous information sources, i.e., things that produce enormous non-structured and or semi-structured data. For this purpose, IoT requires collecting, accessing, processing, and sharing huge amount of data. On other hand, a cloud is capable of providing unlimited storage on demand at low cost and the data stored can be accessed and visualized from anyplace through APIs. Hence for storing the data generated by IoT cloud is the most efficient and cost-effective method.

Computation power The processing capabilities of devices used in IoT are limited. Collected data from various sensors are sent to powerful nodes for aggregation and processing. IoTs computational needs are addressed by the use of on demand model of cloud. IoT with the avail of cloud can perform authentic-time processing of data and thus highly responsive applications can be addressed.

Communication devices Devices with IP address can communicate with each other through a dedicated set of hardware in IoT. Connecting, tracking, and managing devices is made easy by integrating IoT with cloud computing. Any device with an IP address can be accessed over internet from anywhere. IoT controls things in real time through built in applications.

Scalability When cloud integration is provided, any number of things can be integrated or abstracted from the system. Cloud provides scalable approach to IoT. The resources can be varied dynamically according to the need or demand. In cloud, the resources are allocated in accordance with things and applications requirement.

Availability With cloud integration anytime and anywhere, availability of resources becomes effortless. The end user receives continuous services, as the applications connected to the cloud are always up and are running continuously.

Interoperability Heterogeneous devices are used in IoT. Due to different hardware and software configurations compatibility issues may arise. In an IoT environment, maintaining interoperability becomes hard. Cloud solves this problem by providing a communal platform, where all the devices connected can interact with each other, i.e., they can share and exchange data in their compatible data format [3].

3 Infrastructure of a Smart Parking System

The basic infrastructure of a smart parking system is given in Fig. 1.

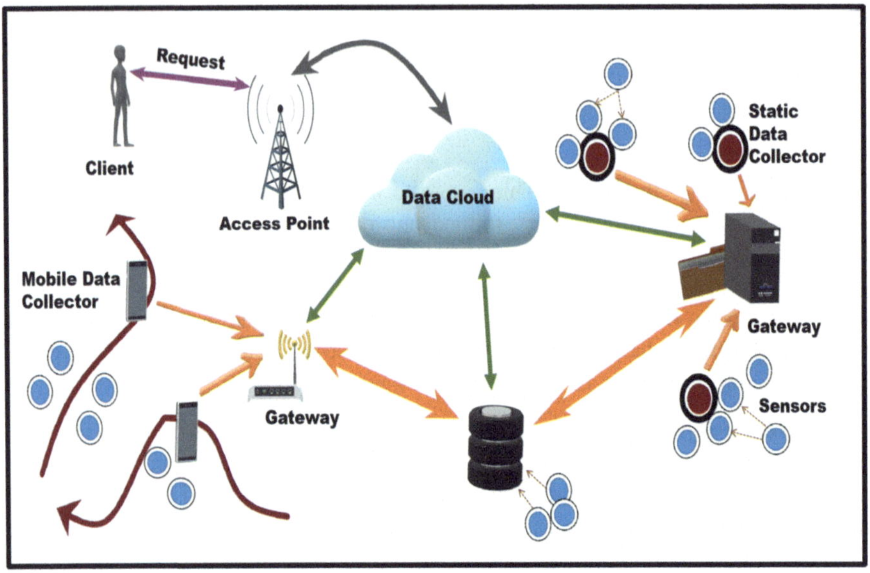

Fig. 1 Architecture model of IoT sensing

3.1 Detection System

The primary function of a parking system depends on devices like sensors to detect the occupancy of the parking slot. Two types of detection/sensing technology namely, vision and sensor-based are predominantly used.

Vision-based methods employ CCTV where generally a single camera visualizes many parking slots and the captured images are sent to a processing software which identifies the state of the parking slot based on comparison method employed in machine vision. In sensor-based method, each parking slot has individual sensor for monitoring. The signals from the sensors are transmitted to the processor for processing through a suitable communicating network.

3.1.1 Vision-Based Method

It is an efficient and cost-effective technology that guides cars to unfilled spaces thereby effectively utilizing the parking spaces. Vision-based detection/sensing technology can be divided into two families.

In the primary category, number of vehicles arriving and departing the parking lot is estimated by the detection system, which will reveal the remaining unfilled slots. In the secondary category, the state of each single space is monitored and is used to guide a car to an unfilled space. This is an intelligent parking system as it aids the drivers to find the unfilled space without much effort by prompting the exact location of the unfilled space and not just the total number of spaces.

Numerous methods have been developed to determine the state of an individual parking space with the help of surveillance cameras, ultrasonic sensors, etc. installed at each slot. Usually, many sensors are required to monitor the entire parking lot but when the cameras are positioned high enough so that they provide a better coverage with minimum cameras. This is especially helpful in open-air parking lots. However, discovery techniques predicated on cameras and picture preparing experience the ill effects of an absence of exactness and can be influenced by ecological or climate conditions.

Chief problems associated with vision-based parking sensing systems are shadow effects, obstruction in image, low resolution due to dull or bright light, and view distortion. Light-colored cars in daylight background may lead astray the detector software to consider it as an unfilled space; similarly, a dark area can be falsely identified as a dark-colored vehicle thereby directing to an unoccupied empty space. Also when there is a change in lighting, it affects the performance of the detector. This is a major problem in the camera-based detection/sensing method [4, 5]. This led to the usage of sensors instead of cameras.

3.1.2 Smart Parking Systems Based on Computer Vision

Lately, most of the research has been targeted toward the computer vision-based methods. Here in the study includes data acquisition, processing, and analysis of images. Computer vision employs computers to compete with human visual sense which implies that the computers are able to learn, make conjecture, and perform suitable actions depending on its visual input. The ultimate aim of computer vision is to enable the computers to observe a visual data in the form of images or videos and to act upon when there is any change in the visual data. For the most part, the procedure is to examine a couple of edges for every second pursued by sending the information to a chief database, along these lines the client can recover the data about the adjustments in the parking area. Takizawa et al. [6] and Funck et al. [7] in 2004 developed a smart parking system wherein they employed a CCTV to identify the presence/absence of a vehicle in a parking space through pixel spotting. A particular amount of pixels in the grayscale was set as a threshold to distinguish between vehicle occupied and unoccupied states in a parking lot. This type of system is highly reliable and accurate. However, these can be utilized in cases where occupancy data are required. In order to acquire the car-park occupancy data Bong et al. [8] in 2008 developed an integrated approach of image processing algorithms termed as car-occupancy information system (COINS). Due to the use of image processing techniques rather than sensors, made this system cost effective. Figure 2 shows the pictorial working of vision-based smart parking system. In the figure below, C1, C2, C3, C4 are cameras highly mounted for better monitoring [9].

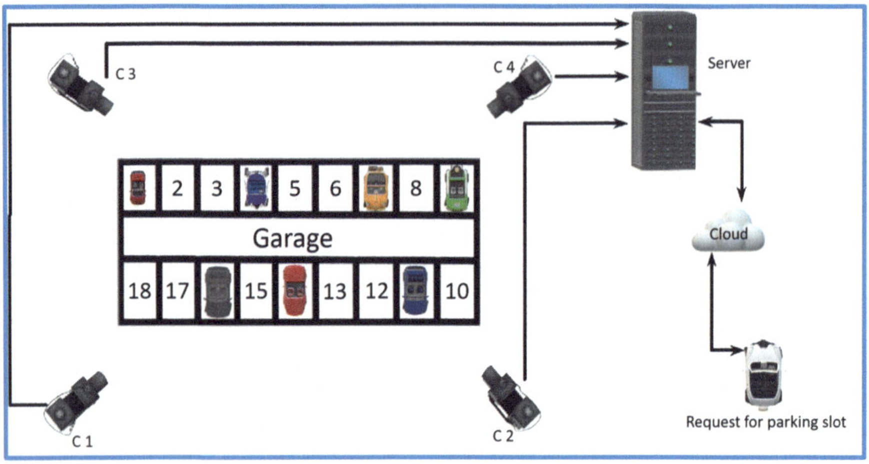

Fig. 2 Pictorial representation of computer vision-based parking system

3.1.3 Sensor-Based Method

Yet another detection method is the usage of sensors to sense the unfilled spaces in a parking lot. Due to the availability of various types of sensing systems, the selection of a suitable sensor depends on factors such as size, reliability, robustness in changing environment, and cost. The sensor technologies can be either categorized as intrusive or non-intrusive.

The intrusive sensors require direct installation on the road surface for which excavation and tunneling has to be done on the road surface. Sensors that come under this category are piezoelectric cables, magnetometers, weight-in-motion sensors, pneumatic tubes, piezoelectric cables, and inductive loops. Wolff et al. [10] used the Earth's magnetic field to discover unfilled parking slots. These sensing devices are directly wired to the control computers.

One circumscription of the intrusive sensor systems is that long and perplexed wiring is required from parking lots to the central control unit. In addition due to the requirement of large amount of sensor systems, the cost of development of such systems is high. Hence, wireless sensors came into play. Tang et al. [11] developed a wireless sensor system using Crossbow Mote products and the extended Crossbow XMesh network architecture. Benson et al. [5] suggested RF transceiver and antenna with an ATMega 128L micro-controller system. Non-intrusive sensors are installed directly on ceilings or ground. These sensors include acoustic and ultrasonic detection systems. They require simpler installations compared to intrusive sensors [12].

3.1.4 Smart Parking Systems Based on RFID Technology

Radio frequency identification (RFID) employs electromagnetic waves to instantaneously identify tagged cars/vehicles. Tags usually store electronic information that transmits data about the object to the reader/antenna through radio waves. This system is still in its prototype stage and is little difficult to realize it practically. In 2007, Pala and Nihat [13] developed an RFID-based parking system that uses a software program to control and report the changes in a parking space. It can also perform tasks such as selecting the nearest unfilled parking space and the location is sent to the driver. Yet another RFID system was proposed by Jian et al. 2008 [14] that contains a Gate-PC controller and embedded Gate hardware. It was a modular parking management platform and can be replaced by any other related system or hardware.

3.2 Processor

A processor is a device that receives input from sensors and processes the input and converts it into a suitable form which is then transferred to the client. A processor has inbuilt memory storage which makes it more efficient in handling numerous data at a single time within microseconds. Figure 3 shows the general working of a processor.

There are many processors, which are suitable for this application, some of them are Infineon, Freescale, Armcortex, auridino, raspberry pi, Tsgate, Tsmote. Due to robustness and ease of configuring, advanced RISC machine (ARM) is being widely used for this application and is discussed below.

In this work, ARM family of generic microprocessors were used. The processor used widely is ARM9TDMI which works on the embedded control from receiving the input from the sensors and transmitting the signals to the users. This processor attributes high performance, compatibility, economical power consumption, and high-speed data transmission for the high-speed IoT applications. This processor fortifies both the 32-bit ARM and 16-bit Thumb ordinant dictation sets, sanctioning the utilizer to trade-off between high performance and high code density which helps to add more features in future. Also it reckons logic to assist in both hardware

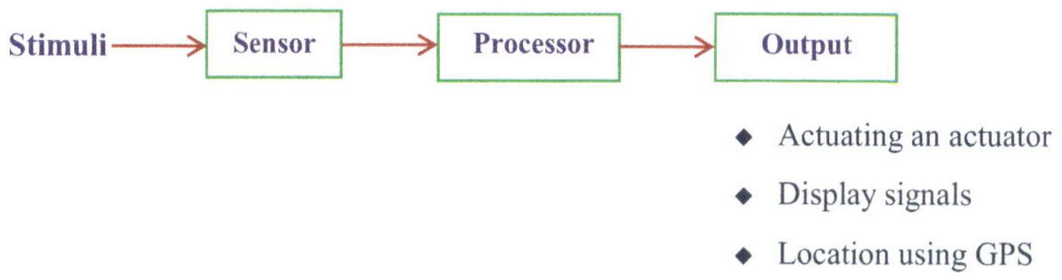

Fig. 3 General working of a processor

and software debugging issues. In addition, the selected processor possesses to add external memory as on the demand required without mutation on the clock speed. The ARM9TDMI also support for co-processors for master slave applications. The ARM9TDMI processor core is constructed with five-stage pipeline consisting of fetch, decode, execute, recollection, and indite stages. The device constructed with simple bus interface affluent connections to either a cached or static RAM-based memory system. An elementary data transfer protocol is catered for master slave support [15].

3.3 Sensor Connectivity

Right after the installation of the sensors on parking spaces, they can form a network and can send out their data to the processor. In order to set up communication between the sensor and the processor, two communication methods are used. They are (1) short range methods such as 802.11ah (Wi-Fi HaLow), Bluetooth/BLE and Zigbee/Z-Wave/DigiMesh, and (2) long range methods such as Sigfox, LoRa, Weightless, Ingenu, and NB-IoT/LTE-M. The long range communication methods have the advantage of utilizing the subsisting radio access network and can communicate with the infrastructures irrespective of place and time. Short range communication methods mostly use wireless sensor networks (WSN) where messages have to be transmitted again through relays. For example, messages have to reach parking meters or other sensors till the last infrastructure in the city. Nowadays, WSN-based smart parking has attracted much attention. However, it suffers from low lifetime, information delay, and wireless link quality limitations [16].

3.3.1 Parking System Based on Wireless Sensor Networks

Since 2005, these kind of systems have been the center of attraction for researchers as they have numerous advantages like rapid sensing, intelligence, speedy deployment, flexibility, and low cost. WSN-based parking systems have been proposed by umpteen researchers. Due to reasonable cost, ease of installation, and configuration, these types of systems are used in monitoring environmental condition with the aid of sensors. This system primarily detects the empty parking space and guides the driver to it through suitable signals. The sensors used in this system are ultrasonic and acoustic based, i.e., they emit sound waves at a particular wavelength which gets reflected back on hitting the object (car or ground) which in turn are converted into suitable visual format by the processor and transmitted to the driver through wireless network. Mathur et al. 2009 [17] in his research proposed equipping the cars with ultrasonic sensors, which drive past the parking spaces to amass occupancy data and upload the data to the centralized database. Once a car sends a request to park, the centralized database provides the necessary information that guides the car to the vacant parking space. Lee et al. 2008 [18] proposed the utilization of an amalgamation of magnetic and ultrasonic sensors for the precise and reliable detection of conveyances in a parking lot, and described a modified version of the min max algorithm for the detection of conveyances utilizing magnetometers.

3.3.2 Smart Parking Systems Based on Vehicle to Infrastructure Communication

Connected vehicle technology (CVT) which depends on wireless data transmission between vehicle and infrastructure (V2I) is a promising technology emerged recently. It proposes an incipient keenly intellective parking technique that depends on developing an incipient VANET-predicated keenly intellective parking to be utilized for keenly intellective steering and perspicacious parking. It refers to vehicle communication system, where the vehicles and the road side devices or units act as the communication nodes, i.e., they can communicate and exchange signals with each other like traffic information, safety warnings, and finding the empty parking slot. This network consists of two types of nodes namely vehicles and roadside station [19], they are assorted as "Dedicated Short Range Communication" devices (DSRC), which works in 5.9 GHz bands under a bandwidth of 75 MHz and has a range of about 1 km. Here two-way communication between the vehicle and infrastructure is possible. In smart parking, the communication with the infrastructure starts with the driver sending their parking request along with driver information (IP), on receiving the request the system confirms the request, allots the paring slot, notifies the driver about the location, and guides the driver to the accurate position of the parking slot. A new parking system called SPARK proposed by Stibor et al. 2007 [20] consists of four parts namely real-time parking navigation, system setting, friendly parking information dispersion, and an intelligent anti-theft protection. In this proposed method, the entire parking lot is under the supervision of three roadside units (RSUs). The RSUs monitor and control the entire parking lot using VANET technology; light sensors are employed for the purpose of detecting empty parking space.

3.3.3 GPS-Assisted Smart Parking System

Using global positioning system (GPS), a vehicle's exact location can be determined. Here, this system is used to determine the empty parking slot and guide the vehicle to that particular location. Chon et al. [21] presented NAPA, a location-based system for smart parking. The server in the system associates buildings on the campus with parking lots in the order of distances to the building. After locating the most proximate available parking lot, the utilizer sends the NAPA server a message that he/she has parked. Another way of smart parking is by booking the available parking slot by sending short messages to the server. First, the free space request must be sent by SMS and the server sends the list of available empty parking space.

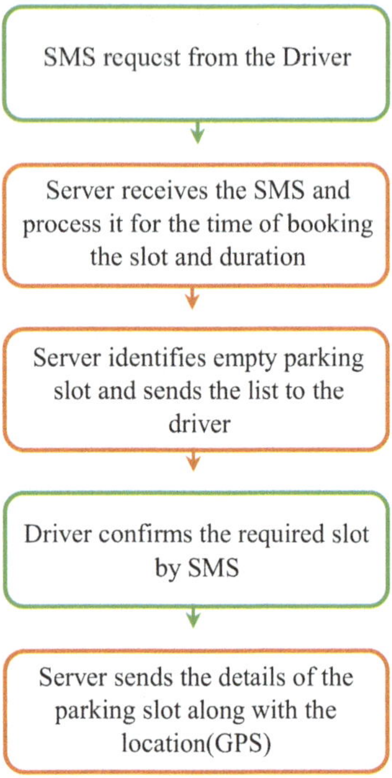

Fig. 4 Flow diagram of SMS-based smart parking system

The booked parking slot location is sent to the user through SMS using GPS [22]. This prototype circuit is simple in design, has high reliability and high accuracy.

Figure 4 shows how the request is generated and how it is completed in a flow diagram. The driver on parking at the location allocated will complete the request cycle by sending an SMS.

3.3.4 Other Hybrid, M2M, IoT Systems

Machine to machine (M2M) IoT systems is exchange of information between two machines without human intervention. It allows direct communication among wireless heterogeneous terminals. There are usually four stages in an M2M communication system. They are (a) data acquisition, (b) transmission of data over a communication network, (c) processing of the data, and (d) result of the information. This property makes M2M more desirable in a smart parking system. The M2M systems always have aggregators that connect multiple end nodes that contain sensors to report the physical changes and a gateway that connects the cellular network for subsequent internet attachment. Certain hybrid techniques were developed by Yeh et al. 2016 [23]; Pham et al. 2015 [24]. One noteworthy hybrid system was developed by Lee et al. in 2016 [25], which detects the unfilled parking spaces through smart sensors and actuators with the middleware copulative clients to terminal contrivances. The system operates via Bluetooth connection between the smartphone and wireless sensor nodes. Another hybrid parking system was proposed by Fraifer and Fernström in 2014 [26] where they developed a prototype using camera nodes and open source computer vision algorithm to identify unfilled parking space. In 2011, Foschini et al. [27] designed and implemented an M2M application that can manage road traffic. It has an integrated IoT retractile pillar management system that restricts vehicular access to particular areas in the city based on standard infrastructure and software components. This is an intelligent parallel technique that uses RFID technology along with gate PC controllers, embedded gate hardware, and two ultrasound range sensors. The vehicle check in and check out is controlled by the RFID tags and reader. This system emerged as a promising application of IoT that can be brought to practical development.

4 Challenges in Implementing IoT

4.1 Technological Challenges

In order to realize a fully functional smart parking system, the following technological constraints have to be overcome.

Scaling of integrated systems In an effective smart parking ecosystem, there are various integrated systems for each area. Each of them may use different hardware and software components, i.e., sensors and communication methods. With the addition of service like electric vehicle charging, the complexity of system integration increases. Therefore to achieve scalability, standard framework like the connected vehicle reference implementation architecture should be used by all the system developers.

Security One of the motives of smart parking systems is to make cashless payments for parking. But this will involve security challenges such as the amenability with payment security regulation, user recognition, and validation or parking occupancy monitoring so as to rightly charge the user. Overall confidentiality of the system has to be guaranteed.

Co-existence of non-smart cars Some of the smart parking systems employ RFID technology for the identification of cars and has RF readers installed in the parking space. In order for the reader to detect an inbound vehicle, the vehicles should possess RF tags. However, it cannot be assumed that all the cars are going to have these tags and thereby this technology cannot be used by those cars. Thus there will be a miscount in the number of cars entering the parking lot and also the occupancy of parking space by these vehicles cannot be detected which sooner or later leads to a chaos in the space provided for parking.

Dynamic pricing models There are a few research and applications that have proposed dynamic pricing models to cut down the time spent in finding the parking slot and overcrowding. In these systems, parking will become more expensive during peak traffic. Also no real-time traffic data is used to estimate the prices. If the real-time traffic data are considered it could be beneficial to efficiently disperse the traffic congestion caused by the drivers searching for a parking spot [28].

5 Conclusion

Owing to increasing vehicle population in urban area and reduced parking space the necessity for advanced parking aid arises. One of the major achievements so for in research is the use of IoT and cloud computing techniques along with sophisticated communication channels. This has reduced the adversity of parking drastically. The smart parking system is classified into three major subsystems; they are gathering information, development of system, and service distribution. Different sensing, monitoring, and processing techniques have been analyzed and discussed in this chapter and made ready for the readers to understand the need and working of the smart parking systems. Engineering insights on how IoT and cloud computing plays a major role in this system is also discussed and the best way to get the highest efficiency of this system is to integrate IoT and cloud computing together with the suitable communication techniques. This chapter helps to understand the working and the required facilities to be considered in the parking spaces to implement smart parking systems. This helps in reducing the trial and error for the implementation thereby reducing its cost. This chapter also highlights the challenges to overcome while implementing a smart parking system. State-of-the-art technology throws out suggestions for further development in research on smart cities by improving the IoT architecture.

References

1. Atif, Y., Ding, J., & Jeusfeld, M. A. (2016). Internet of things approach to cloud-based smart car parking. *Procedia Computer Science, 98*, 193–198.
2. Zanella, A., Bui, N., Castellani, A., Vangelista, L., & Zorzi, M. (2014). Internet of things for smart cities. *IEEE Internet of Things Journal, 1*(1), 22–32.
3. Khanna, A., & Anand, R. (2016, January). IoT based smart parking system. In *2016 International Conference on Internet of Things and Applications (IOTA)* (pp. 266–270). IEEE.
4. Yan, G., Yang, W., Rawat, D. B., & Olariu, S. (2011). SmartParking: A secure and intelligent parking system. *IEEE Intelligent Transportation Systems Magazine, 3*(1), 18–30.
5. Benson, J. P., O'Donovan, T., O'Sullivan, P., Roedig, U., Sreenan, C., Barton, J., Murphy, A., & O'Flynn, B. (2006, November). Car-park management using wireless sensor networks. In *Proceedings. 2006 31st IEEE Conference on Local Computer Networks* (pp. 588–595). IEEE.
6. Takizawa, H., Yamada, K., & Ito, T. (2004). Vehicles detection using sensor fusion. In *Intelligent Vehicles Symposium, 2004 IEEE* (pp. 238–243).
7. Funck, S., Mohler, N., & Oertel, W. (2004). Determining car-park occupancy from single images. In *Intelligent Vehicles Symposium, 2004 IEEE* (pp. 325–328).

8. Bong, D., Ting, K., & Lai, K. (2008). Integrated approach in the design of car park occupancy information system (COINS). *IAENG International Journal of Computer Science, 35*(1), 7–14.
9. Fraifer, M., & Fernström, M. (2016). Investigation of smart parking systems and their technologies. In *Thirty Seventh International Conference on Information Systems. IoT Smart City Challenges Applications (ISCA 2016), Dublin, Ireland* (pp. 1–14).
10. Wolff, J., Heuer, T., Gao, H., Weinmann, M., Voit, S., & Hartmann, U. (2006, September). Parking monitor system based on magnetic field sensor. In *2006 IEEE Intelligent Transportation Systems Conference* (pp. 1275–1279). IEEE.
11. Tang, V. W., Zheng, Y., & Cao, J. (2006, August). An intelligent car park management system based on wireless sensor networks. In *2006 First International Symposium on Pervasive Computing and Applications* (pp. 65–70). IEEE.
12. Kianpisheh, A., Mustaffa, N., Limtrairut, P., & Keikhosrokiani, P. (2012). Smart parking system (SPS) architecture using ultrasonic detector. *International Journal of Software Engineering and Its Applications, 6*(3), 55–58.
13. Pala, Z., & Inanc, N. (2007, September). Smart parking applications using RFID technology. In *2007 1st Annual RFID Eurasia* (pp. 1–3). IEEE.
14. Jian, M. S., Yang, K. S., & Lee, C. L. (2008). Modular RFID parking management system based on existed gate system integration. *WSEAS Transactions on Systems, 7*, 706–716.
15. *ARM9TDMI Technical Reference Manual*. Retrieved from http://infocenter.arm.com/help/topic/com.arm.doc.ddi0180a/DDI0180.pdf
16. Lin, T., Rivano, H., & Le Mouël, F. (2017). A survey of smart parking solutions. *IEEE Transactions on Intelligent Transportation Systems, 18*(12), 3229–3253.
17. Mathur, S., Kaul, S., Gruteser, M., & Trappe, W. (2009, May). ParkNet: A mobile sensor network for harvesting real time vehicular parking information. In *Proceedings of the 2009 MobiHoc S 3 Workshop on MobiHoc S 3* (pp. 25–28). ACM.
18. Lee, S., Yoon, D., & Ghosh, A. (2008, May). Intelligent parking lot application using wireless sensor networks. In *CTS* (pp. 48–57).
19. Geng, Y., & Cassandras, C. G. (2012). A new "smart parking" system infrastructure and implementation. *Procedia-Social and Behavioral Sciences, 54*, 1278–1287.
20. Stibor, L., Zang, Y., & Reumerman, H. J. (2007, March). Evaluation of communication distance of broadcast messages in a vehicular ad-hoc network using IEEE 802.11 p. In *2007 IEEE Wireless Communications and Networking Conference* (pp. 254–257). IEEE.
21. Chon, H. D., Agrawal, D., & El Abbadi, A. (2002, February). NAPA: Nearest available parking lot application. In *Proceedings 18th International Conference on Data Engineering* (pp. 496–497). IEEE.
22. Hanif, N. H. H. M., Badiozaman, M. H., & Daud, H. (2010, June). Smart parking reservation system using short message services (SMS). In *2010 International Conference on Intelligent and Advanced Systems* (pp. 1–5). IEEE.
23. Yeh, H. T., Chen, B. C., & Wang, B. X. (2016). A city parking integration system combined with cloud computing technologies and smart mobile devices. *Eurasia Journal of Mathematics, Science & Technology Education, 12*(5), 1231–1242.
24. Pham, T. N., Tsai, M. F., Nguyen, D. B., Dow, C. R., & Deng, D. J. (2015). A cloud-based smart-parking system based on Internet-of-Things technologies. *IEEE Access, 3*, 1581–1591.
25. Lee, C., Han, Y., Jeon, S., Seo, D., & Jung, I. (2016, January). Smart parking system for Internet of Things. In *2016 IEEE International Conference on Consumer Electronics (ICCE)* (pp. 263–264). IEEE.
26. Fraifer, M., & Fernström, M. (2015). Smart car parking system prototype utilizing CCTVs. Alliance, Postgraduate Res. Day. *Computer Science and Information Technology*, 7.
27. Foschini, L., Taleb, T., Corradi, A., & Bottazzi, D. (2011). M2M-based metropolitan platform for IMS-enabled road traffic management in IoT. *IEEE Communications Magazine, 49*(11), 50–57.
28. Sauras-Perez, P., Gil, A., & Taiber, J. (2014, November). ParkinGain: Toward a smart parking application with value-added services integration. In *2014 International Conference on Connected Vehicles and Expo* (ICCVE) (pp. 144–148). IEEE.

Mathan Chandran is currently pursuing his doctorate program in PSG College of Technology, Coimbatore, Tamilnadu, India. He completed his master's degree in automotive engineering from the same institute. He is receiving CSIR SRF Fellowship and has published two journal papers. His research interests include PEM fuel cell, fuel cell vehicle development, high pressure hydrogen storage, IoT for sustainable development, and product development for social-economic problems.

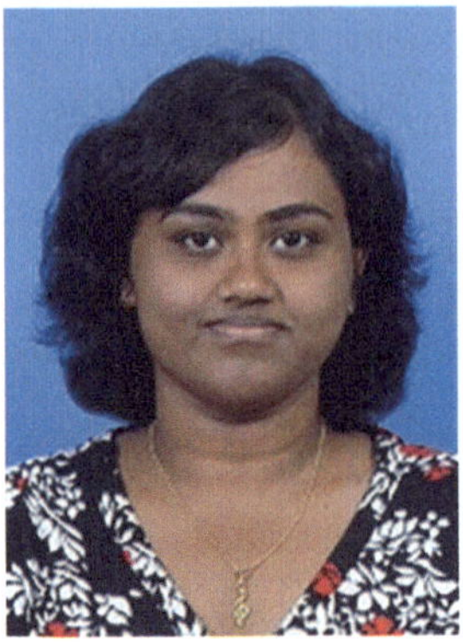

Susan Immanuel is working as a JRF in DRDO sponsored project and currently pursuing her doctorate in 2D nanomaterials for electrochemical sensor application. She has completed her master's in nanoscience and technology from PSG College of Technology, Coimbatore, Tamil Nadu, India. She has published over six international journal papers of high quality. Her research interests include NADH biosensor, cholesterol nanobiosensors, dopamine nanobiosensors, biomaterials, tissue engineering and nanoparticles for drug delivery, biological fuel cell, and environmental-friendly technological development.

Thanarajan Kumaresan is currently a senior research fellow in DST project at PSG College of Technology, Coimbatore, India. He received his DAE (2011) from Vetri Vinayaha Polytechnic College Trichy, BE (2014) from Tamilnadu College of Engineering Coimbatore, and ME (2017) from PSG College of Technology, Coimbatore, Tamilnadu, India. He has all his degrees in automobile engineering and now pursuing PhD in fuel cells technology at PSG College of Technology, Coimbatore, India. He has published four international journal papers of high quality. He is a member in American Society of Mechanical Engineers (ASME). His research interests are fuel cells, sustainable development technologies, and automobile power train.

R. Suresh Kumar is currently working as an assistant professor in the Department of Automobile Engineering at PSG College of Technology, Coimbatore. He has graduated in masters in automotive engineering from the same institution. His thrust area is aging characteristics estimation of lithium-ion batteries for hybrid and electric vehicle applications.

K. Vishnu Murthy is currently working as an assistant professor in the Department of Electrical and Electronics Engineering at Sri Krishna College of Technology, Coimbatore. He has graduated his master's degree in Anna University, Coimbatore. He has published more than five research articles in peer reviewed journal. His research interest includes electrical drives, IoT system application in electrical drives and control and data science.

Implementing Cryptographic Internet of Things in Patient-Centered Network to Vestige Alzheimer and Dementia Patient

M. Kowsigan, S. Ramamoorthy, Sibi Amaran, and D. Jeyabharathi

1 Introduction

Alzheimer's disease is a chronic neurodegenerative disease and the most common type of dementia which affects mostly elderly people resulting in progressive loss of memory, deficiency in social interaction, and impairment of daily activities. The symptom for Alzheimer's which is most commonly experienced by the affected people is short-term memory loss, which gradually grows and renders them unable to remember even the identities of their own family members and in some cases, their own details. It is expected that the burden caused by dementia in developing countries will follow an upward trend due to an increase in longevity and prevalence of risk factors such as hypertension [1]. According to research, most people currently affected with any sort of dementia have not undergone a formal diagnosis. In high-income countries, it is noted that only 20–50% of dementia cases are recognized and documented [2]. The World Health Organisation (WHO) has predicted that by the year 2025, about 75% of the estimated 1.2 billion people aged 60 years and older will reside in developing countries. The cost of health care for Alzheimer patients is huge as extreme care has to be taken while treating such neuro-degeneric diseases. The proposed project is to design and implement a system to track and locate Alzheimer's disease patients who are prone to memory

M. Kowsigan (✉) · S. Ramamoorthy
Department of Computer Science and Engineering, School of Computing, SRM Institute of Science and Technology, Chennai, India

S. Amaran
Department of Computer Science and Engineering, SRM Institute of Science and Technology, Kattankulathur, India

D. Jeyabharathi
Department of Information Technology, Sri Krishna College of Technology, Coimbatore, India

G. R. Kanagachidambaresan et al. (eds.), *Internet of Things in Smart Technologies for Sustainable Urban Development*, EAI/Springer Innovations in Communication and Computing, https://doi.org/10.1007/978-3-030-34328-6_11

loss and suffer from lapses of concentration [3]. In such critical circumstances, tracking the movements of the affected patient is beneficial indeed as it helps us to know the exact location of the patient and predict their condition. In this chapter, the patient carries a tracking module which when queried by the caregiver, automatically locates itself via a GPS receiver [4] and then relays the location to the caregiver via GSM. This system is very efficient and highly useful in tracking the patient's movement. Mobile Hub software is capable to run almost every Bluetooth-enabled and Android-based smartphone which can be used by the caregiver to locate the patient whenever he/she wants to. In case of a sudden and unexpected panic situation, an alarm can be activated either manually by the patient or automatically by the accelerometer sensor in the smartphone. When this alarm signal is given, the central dispatcher is able to acquire the location based on the GSM or GPRS cell information immediately and the information is sent to the cloud using IOT via the SIM module. This allows the concerned people to view the current and exact location of the patient and take necessary action. In case of any abnormality in the patient's health, the concerned people will be notified about the patient's condition either through an SMS or a phone call. The implementation of this system can be done using SIM 808. A major issue faced with the location tracking is the security of the data. This can be controlled by implementing data encryption on the location data which will ensure the effectiveness and security of the information transmitted/received. A security algorithm is a mathematical procedure used to encrypt data which requires the help of a software key once encrypted in the form of a cipher text, in order to convert the text back to its normal form. In this project, enhanced RSA (Rivest-Shamir-Adleman) algorithm is used to encrypt the data. This algorithm uses a public key as well as a private key with signature along with an Arduino Pro Mini. The main objective of this project is to provide a reliable and efficient real-time health-monitoring and location-tracking system for dementia patients and also eliminate the need for expensive facilities.

2 Literature Review

The shoe is a wearable device which is used for the person affected with dementia [5]. There are totally seven stages in dementia, the most difficult stage is from stage 1 to stage 2 because the physician himself will not be able to find out the transformation of the patient and the caregivers [6]. The shoe embedded with GPS tracking system provides path guidance [7]. The AD patient will be reminded about his appointments and meetings via SMS [8]. The components used here to design the smart shoe are GPS module, WiFi/GSM, cloud, and Arduino. A wearable Steps is a pair of shoes designed for those suffering from AD. Given AD patients' propensity to forget and lose what they are carrying, we embed technologies within the shoe as a daily fashion item that patients wear when they go out for a walk. The shoe supports two functions: they alert patients who fall into a mental trance and they monitor the patients' location for their caregivers. This is the era of wearable technology which will be the future of human science [8]. The shoe is water resistant and it will not be damaged if at all it is immersed in water. The battery of the pocket is "piezoelectric" so that the battery doesn't get drained off and it produces energy as and when the patient starts walking [9]. No medicines have been found for the disease dementia; even today researchers are struggling to find a proper medicine for the people suffering [10]. So this shoe might be very helpful for them and for their caregivers to make them stable and send an alert of what they are up to and what kind of remedial measures they can take to cure them [11]. The systems have also been implemented in bus tracking and monitoring locations in an efficient way. Version control and bug tracking systems comprise of large measures of ancient data that can provide deep insight into the evolution of the internet of things [12]. GPS is constantly turning into universal utility as the expense of installation of the technology into automobiles, computer, machinery, cellular phones decreases. GPS also have been advanced into miniature devices for further applications in other fields of sciences. Molding the world of science through mammoth networks of wireless network ports, specifically for implementations like environmental monitoring of water and soil, need these nodes to be very minute, lightweight, and unobtrusive. Real world considerations such as the miniature size, expense, and power constraints of ports preclude the power of reliance on GPS of all ports in these interconnected networks [13]. The modified approach which is an enhancement over traditional RSA algorithm may include prime numbers, public keys, K-NN algorithm, and exponential powers both for the sender and receiver [14].

3 Proposed System

Figure 1 explains the components used in the SIM integrated cloud model include the Arduino IDE, Cloud Server, SIM 808.

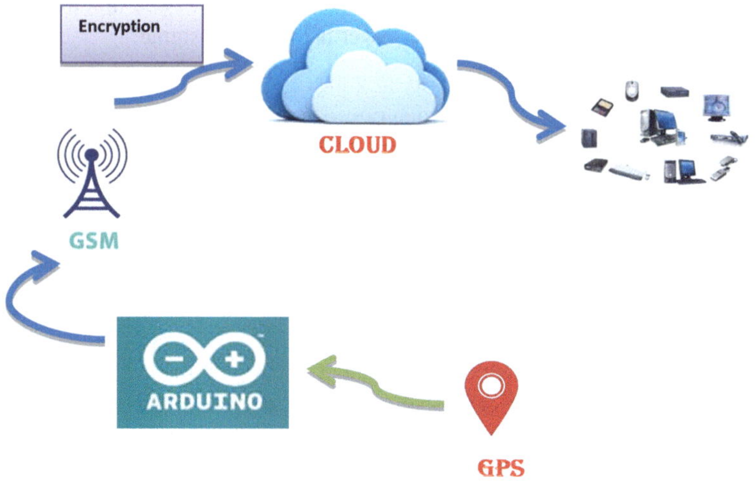

Fig. 1 Working model

- At first download the Arduino IDE into your computer which is the software used to embed C programming into the Arduino board for application. Create a channel consisting of two fields where one of them is meant to show the latitude and the other for longitude.
- The next step includes setting up of components in place such as placement of the Arduino and SIM 808 into the Pocket. Update the write and the read key from the channel to the code and upload the results in the cloud after creating a channel along with introduction of data variables [15].

The tools used for development include Arduino Pro Mini, SIM 808 Module, and Arduino IDE along with cloud server. Arduino Pro Mini is used to get the readings from SIM808 and perform actions by sending certain AT commands.

The input data from the source are secured using RSA algorithm in the Arduino Pro Mini which provides a secured cipher Text. In order to enhance protection while transmitting data we proposed security algorithm in the Channel path. This encrypted data is uploaded to the cloud. The data can be retrieved from the cloud and original data is obtained from SIM 808.

SIM 808 is used to get the location and transmit the data through GPRS signal to the cloud. Arduino IDE is used to program the Arduino Board and send commands to view the output in text format. The cloud server is used as major cloud storage for further analysis and virtualization. The development language to be used as the platform for the coding part is made more efficient with the usage of Embedded C.

The cloud server is the IOT [16] analytic server used to store the coordinate values to the server. The location data includes the latitude and the longitude values that can be viewed as a graph by the analyzer. The coordinates of the location is sent to MATLAB visualization to show the location of the patient in Google Maps. Channel Id and read key are used to read the data from the server.

The following steps are used to deploy the cloud environment:

- Send the AT commands to the SIM 808 module through Arduino Pro Mini.
- Get the location coordination of the patients from the SIM 808 module.
- Send the location coordinates to the server using GPRS signal.
- View the data from web server.

Figure 2 describes the flow chat diagram describing how the system gets their input and how they are processing data and the output of the system.

The above flow chart (Fig. 2) describes the cloud storage is an exemplary of data that is being binned in which the digital data is stockpiled in coherent pools and the physical storage intervals multiple servers and the physical surrounding is typically owned and coordinated by a hosting company. These cloud storage providers are responsible for keeping the data available and accessible, and the physical environment protected and running.

Storage and retrieval is enabled and made efficient with the help of cloud server protocol. Cloud serves as an extensive virtual storage for massive databases. The storage is categorized according to the features for easy retrieval. Every application featured with a backend needs standard and reliable cloud service [13]. There are

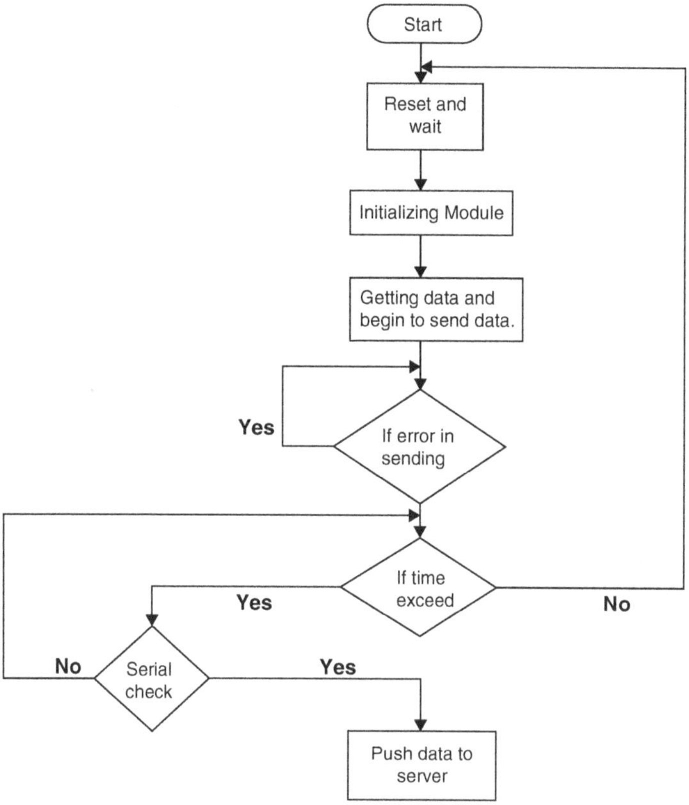

Fig. 2 Flow chart processing data

paid and unpaid types of clouds hired according to the individual's requirements. Our project is concerned to mammoth data which would need a wider platform for storage and retrieval. Though our data is not confidential yet it has to be stored in a cloud service to process further and analyze the data. The GSM is connected to the cloud via a router and instantaneous location of the patient is transferred and stacked into the cloud service. As a cloud user, one would expect best in class performance and best pricing for all his storage needs. GPS or global positioning system is a set of interconnected orbiting satellites revolving around the planet earth that usually sends precise details of their current position on earth. The signals are thus extracted by GPS receivers called the navigation devices. They are used to calculate the pin-point position, velocity and time at the vehicles exact area of location.

GPS is popular for its defense purposes and was first developed by the Americans to aid in its worldwide intelligence efforts at the peak of the Cold War. Ever since the past few decades, the GPS has been available at ease to anyone provided with a tracking receiver. Airlines, trucking firms, shipping industries, and organizations irrespective of the location use the tracking system to trace automobiles, follow the apt to get them from A to B in the shortest possible time.

People and other industrial firms hire storage capacity from the providers to store user data, industrial data, or application data. Cloud storage is unified object storage from live data to data archiving. With low complexity and consistent API across massive storage it introduces a storage you can use. Gone are the days of retrieving data and tape backup. GSM device has gotten the most extra-limited applications and finds its place in almost all appliances relating to IoT [12]. It is a cell phone standard and used internationally. GSM originally stood for group special mobile but now it had been changed to global system for mobile communications time while. We are here to efficiently use global mobile communications chip in the most possibly extensive way thus paving way to a new generation innovation which can be used for tracking with minimal efforts and is a child's play for new users even. We are hereby placing the GSM into the pocket to trace the exact location of the person. The GSM also provides live screening of the location on the rear end by the user using Google Maps. These map navigations are useful to detect the rate of change of position of the person being tracked. Arduino serves as an open source electronic platform via which one can access hardware and software easily. The Arduino is known for its flexibility to blend with any kind of IoT setup and has got unique features. We use Arduino to switch on a sensor or to finger on a button. Since decades Arduino has been serving as the brain of millions of major projects making them easier to use due to its compatibility. A worldwide community of professionals have been gathering this open source platform for achievements in technology. Cardiac care assistance using self-configured sensor network—a remote patient monitoring system was used for this research [17].

Efficient scheduling in computational grid with an Improved Ant Colony Algorithm was used to schedule the database of the patients in an efficient manner [18]. An efficient performance evaluation model for the resource clusters in cloud environment using continuous time Markov chain and Poisson process was used in this research to identify the performance of the tracking system [19]. A novel resource clustering model to develop an efficient wireless personal cloud environment was to cluster the tracking resources to enhance its performance [20]. Markovian model based trustworthy architecture was taken as reference to build the proposed system architecture [21].

The patient suffering from Alzheimer has to be tracked and monitored for their safety by their caregivers. The exact location of the person is being made available to the caregivers via a GSM module. We place the module into the pocket of the patient without their knowledge. Continuous monitoring of the patient is done. When the patient leaves his home he might forget his wallet or phone or any other accessory but it is 100% probable that they would be wearing his pocket while stepping out. The pocket would be fabricated with a very fine layer of chip embedded sole which would never get to the patient's notice [22]. The pocket is embedded with a GPS and an Arduino and a battery thus the GPS is connected to the cloud with the help of GSM and Arduino and the location and time are stored in the cloud database and monitored in your android phone.

Cryptography plays a major role in data protection when it is transmitted from one component to another or cloud. So while processing the data the intruders may affect our system. In order to overcome these issues, we introduced RSA algorithm [6] which solves an unapproachable challenge in network security, enabling the security but exchange of encrypted communication between the component and the cloud.

The following is the description of how RSA works with the Arduino board:

- The RSA is the cipher where plaintext and cipher text are integers.
- The size used in RSA [7] is 128 bits.
- Key pair function is enabled for public and private exponent.
- By using encryption and decryption steps the cipher text is converted to original message which is enclosed with signature.

4 Results and Discussion

In the proposed system, GPS tracker sends the location information to the Arduino board. By implementing, RSA algorithms in the Arduino board the location values are encrypted for the security and the encrypted values are sent to the Matlab and it shows the values in the graphical representation for our experimental purpose [15]. The encrypted location values are sent to cloud storage using GSM module. Figures 3 and 4 show the longitude and latitude values in Matlab. In Figs. 3 and 4, X-axis represents the time period and Y-axis represents coordination of longitude and latitude, respectively.

Figure 5 shows how the current location of the patient in terms of latitude and longitude is converted to cipher text using RSA algorithm [8] with signature before transmitting from the cloud and when it is coming to the user it is again accurately converted to plain text and the latitude and longitude is restored.

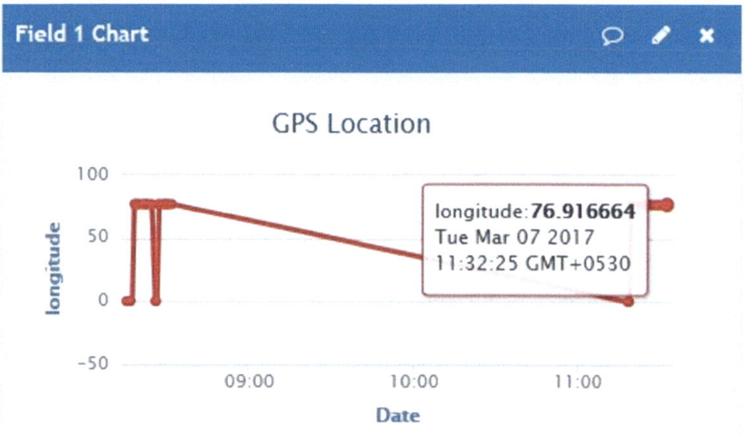

Fig. 3 Longitude location

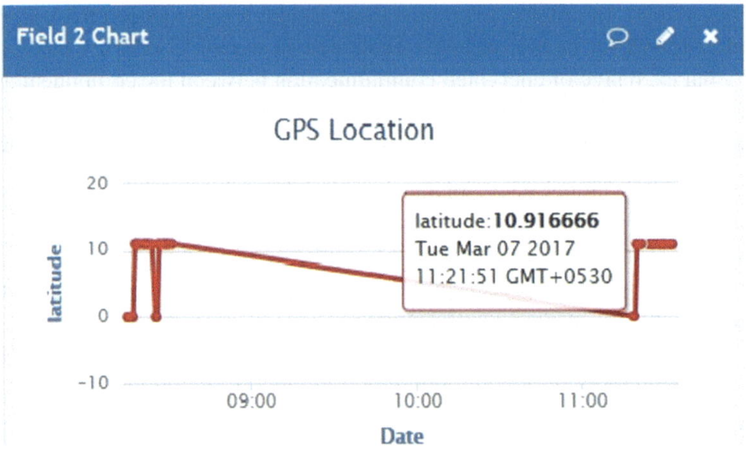

Fig. 4 Latitude location

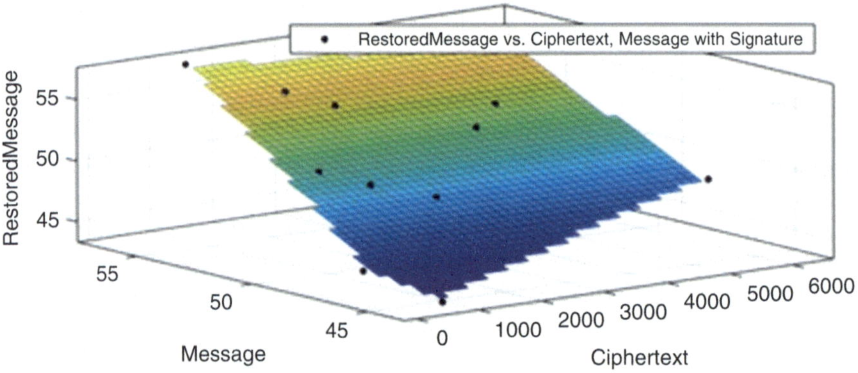

Fig. 5 Encryption and decryption of sensitive data

Figure 6 shows the current location of the patient using GPS by using the longitude and latitude values, [22] and we can see the patient's current location in our mobile phone through Google Maps. Safety zone was created by caretakers, so that when the patient crossed the safety zone, the alert message will be provided to the caretakers. By using the message, the caretakers can take the appropriate action. The safety zone is the where patient can remain safe. When they move out from that zone it seems danger for the patient. Safety zone will be in the range of 500 m around the place where the patients are cared.

The error rate is calculated to find the percentage of accuracy in the data that could be received at the other end. If we send data as plain text it will be modified by the intruders. To prevent this, we encrypt the data and send. Figure 7 shows that the percentage of error in the data transferred without encryption is greater than the encrypted data.

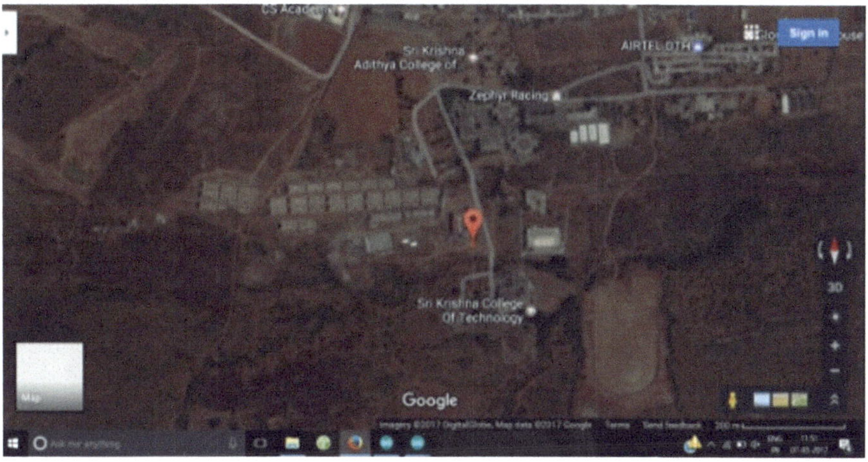

Fig. 6 Location of the patient

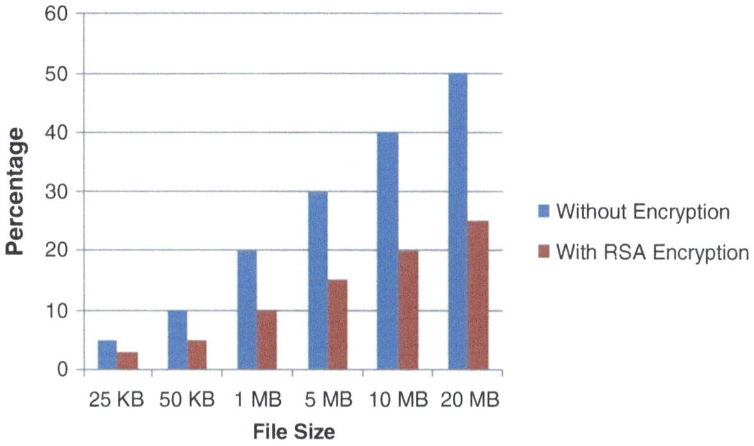

Fig. 7 Error rate vs data with and without encryption

The execution time is defined as the time taken to run the code. The difference between the time taken to send the normal data and the encrypted data is measured in the Fig. 8. It shows that the execution time of encrypted data takes relatively more time compared to the time taken by the normal data.

The data should be received in decrypted form so in Fig. 9 the time taken to send normal data and decrypted data is shown .It explains that the execution time of decrypted data takes relatively more time compared to the time taken by the normal data.

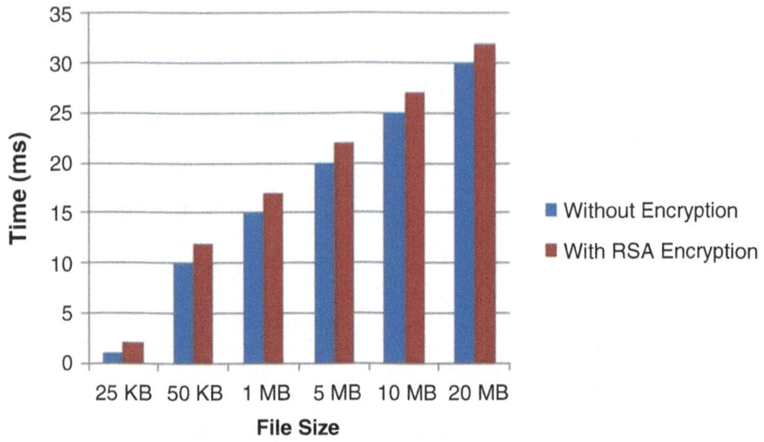

Fig. 8 Execution time of normal data vs encrypted data

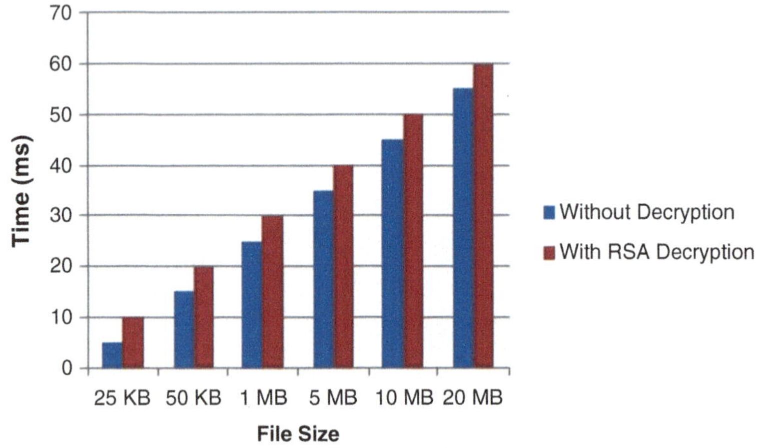

Fig. 9 Execution time of normal data vs decrypted data

5 Conclusion

Nowadays there is no special device to track the location of the Alzheimer's patient and sending the location to the cloud with security through IOT. The main problem that all are trying to solve in IOT is data security. Attackers can easily attack the data that are sent through IOT. But the proposed system can solve the data security problem by using RSA algorithm and send the secured location data to the cloud. By using secured location data, caretaker can monitor the Alzheimer patient's location during their last stages of the disease and take the necessary action for it.

References

1. Kalaria, R. N., Maestre, G. E., Arizaga, R., Friedland, R. P., Galasko, D., Hall, K., et al. (2008). Alzheimer's disease and vascular dementia in developing countries: Prevalence, management and risk factors. *Lancet Neurology, 7*, 812–826. [PUBMED].
2. WHO. (2002). *Active aging: A policy framework. 2002 health report*. Geneva: World Health Organization.
3. Mhamdi, J., & El Abkari, S. Contriving an RFID system for Alzheimer patients tracking. In *2015 Third International Workshop on RFID And Adaptive Wireless Sensor Networks (RAWSN)*, Agadir, Morocco, May 13–15, 2015.
4. Paiva, S., & Abreu, C. (2012). Low cost GPS tracking for the elderly and Alzheimer patients. *Procedia Technology, 5*, 793–802, Elsevier Ltd. HCIST 2012—International Conference on Health and Social Care Information Systems and Technologies.

Theses

5. Dubois, B., Feldman, H. H., Jacova, C., Hampel, H., Molinuevo, J. L., Blennow, K., et al. (2014). Advancing research diagnostic criteria for Alzheimer's disease: The IWG-2 criteria. *Lancet Neurol, 13*(6), 614–629.

International Conference

6. Nema, M., Stalin, S., & Tiwari, R. (2016). RSA algorithm based encryption on secure intelligent traffic system for VANET using Wi-Fi IEEE 802.11p. In *2015 International Conference on Computer, Communication and Control (IC4), 2015 IEEE explorer*, Indore, India.

International Journal

7. Meneses, F., Torres, J., Miranda, A., & Nuela, D. (2016). RSA encryption algorithm optimization to improve performance and security level of network messages. *IJCSNS International Journal of Computer Science and Network Security, 16*(8), 55–62.
8. Singh, S. K., Manjhi, P. K., & Tiwari, R. K. (2016). Data security using RSA algorithm in cloud computing. *IJARCCE, 5*(8), 11–16. https://doi.org/10.17148/IJARCCE.2016.5803.
9. Chepuri, S. (2017). An RGB image encryption using RSA algorithm. *International Journal of Current Trends in Engineering & Research (IJCTER), 3*(3), 1–7.
10. Jagadiswarya, D., & Saraswadya, D. (2016). Biometric authentication using fused multimodal biometric. *Procedia Computer Science, 85*, 109–116.
11. Khambra, D., & Dabas, P. (2017). Internet of things: A perspective from security and privacy. *International Journal of Advanced Research in Computer Science, 8*(5).
12. Hebbar, U., Sultana, S., & Sujith, A. Assessment of security challenges and remedies in IoT applications. *International Journal of Science, Engineering and Technology*. ISSN (O): 2348–4098, ISSN (P): 2395–4752.
13. Tedeschi, S., & Roy, R. (2017). Secure IoT devices for the maintenance of machine tools. *Procedia CIRP, 59*, 150–155.

14. Kanmani, R., & Basha, A. J. (2015). Implementing automized E-ticket using RFID in IOT. *Indian Journal of Emerging Electronics in Computer Communication, 2*(1), 366–371.
15. Kowsigan, M., Rajkumar, S., Seenivasan, P., & Kumar, C. V. (2017). An enhanced job scheduling in cloud environment using improved metaheuristic approach. *International Journal of Engineering Research & Technology, 6*(2), 184–188.
16. Safi, A. (2017). Improving the security of internet of things using encryption algorithms. *World Academy of Science, Engineering and Technology International Journal of Computer, Electrical, Automation, Control and Information Engineering, 11*(5).
17. Dhulipala, V. R. S., & Kanagachidambaresan, G. R. (2014). Cardiac care assistance using self configured sensor network—a remote patient monitoring system. *Journal of the Institution of Engineers (India): Series B, 95*(2), 101–106.
18. Rajeshkumar, J., & Kowsigan, M. (2011). Efficient scheduling in computational grid with an improved ant colony algorithm. *International Journal of Computer Science and Technology, 2*(4), 317–321.
19. Kowsigan, M., & Balasubramanie, P. (2018). An efficient performance evaluation model for the resource clusters in cloud environment using continuous time Markov chain and Poisson process. *Cluster Computing*. https://doi.org/10.1007/s10586-017-1640-7.
20. Kowsigan, M., & Balasubramanie, P. (2019). A novel resource clustering model to develop an efficient wireless personal cloud environment. *Turkish Journal of Electrical Engineering & Computer Sciences, 27*(3), 2156–2169.
21. Kanagachidambaresan, G. R., Dhulipala, V. R. S., & Udhaya, M. S. (2012). Markovian model based trustworthy architecture. *Procedia Engineering, 30*, 718–725.
22. Kowsigan, M., Kalicharan, S., Karthik, P., Manikandan, A., & Manikandan, R. (2017). An enhanced job scheduling in cloud environment using probability distribution. *International Journal of Engineering and Technology, 9*(2), 1374–1381.

Advanced Energy Management System for Smart City Application Using the IoT

A. Sampathkumar, S. Murugan, M. Sivaram, Vishnu Sharma, K. Venkatachalam, and Manikandan Kalimuthu

1 Introduction

About 84% of the world's population has access to electricity, and many people have gained access to electricity in the past decade. Nowadays, the undertaking of ideally planning for a vital and progressive society is becoming increasingly troublesome. The electricity arrangement framework that endeavors the tasks of transmission, circulation and the executives is called a matrix. The smart grid (SG) can partner with and improve the old power grid framework [1]. It joins 2-way correspondence innovations, smart meters, great machines and RES to help convey the products of a framework that is dependable, self-recuperating, cost-proficient, secure and progressively reasonable. A savvy home is a vital unit of a good matrix. The idea of a smart home was first developed in the mid-1990s [2]. With the increase of electricity use, the need to help the general effectiveness of electrical frameworks is further expanding. Subsequently, vitality productivity is changing into an increasingly

A. Sampathkumar (✉) · K. Venkatachalam · M. Kalimuthu
School of Computing Science and Engineering, VIT Bhopal University, Bhopal, India
e-mail: sampath.kumar@vitbhopal.ac.in

S. Murugan
Department of Computer Science and Engineering, Mewar University, Chittorgarh, India
e-mail: smartrugans@gmail.com

M. Sivaram
Department of CSE, Lebanese French University, Erbil, Kurdistan Region, Iraq
e-mail: sivaram.murugan@lfu.edu.krd

V. Sharma
Department of CSE, Galgotias College of Engineering and Technology, Greater Noida, Uttar Pradesh, India
e-mail: vishnu.sharma@galgotiacollege.edu

G. R. Kanagachidambaresan et al. (eds.), *Internet of Things in Smart Technologies for Sustainable Urban Development*, EAI/Springer Innovations in Communication and Computing, https://doi.org/10.1007/978-3-030-34328-6_12

troublesome errand for every SG and smart homes. The EMS streamlines, and the executives deal with the tasks of transmission, circulation and the board. Two significant elements of EMS are the request side administration (DSM) and supply side administration (SSM). DSM is a procedure proposed by utilities, which causes electricity buyers to change their electricity utilization designs through training.

As a rule, the point of DSM is to teach the vitality customer to utilize less electricity throughout on pinnacle hours and satisfy their vitality necessities during off pinnacle hours. Then again, SSM deals with various vitality age assets and improves the responsibleness of offers. Multi-Agent System (MAS) innovation assumes a horrendous potential job in SG and smart homes. In writing, it has been utilized for the point of reproductions, request reaction, activity, SG control, control of microgrid, stockpiling frameworks, smart home administration and for administration reclamation [3]. In MAS, various specialists speak with each other to accomplish a specific objective in an exceptionally viable way. This grants operators the ability to talk, associate and conference with each other for the practical utilization of vitality assets [4]. In this examination work, HEMS is implemented to exploit MAS for smart homes. A partner algorithmic standard intends to create a harmony among interests and give smart machines a customary zone unit utilizing a swarm insight improvement system, BPSO, to help understand the most straightforward calendar for each apparatus [5]. This technique coordinates energy storage system (ESS) that consolidates work unit, turbine (WT), stockpiling framework and principle network (MG). This strategy permits need strategies to blend with ESS. Time of utilization (TOU) valuation duty is utilized that has low, medium and high top hours.

2 Related Works

Currently, around 55% of the world population lives in urban areas [6]. Subsequently, progress in organization is imperative to improve electricity use in urban regions (for instance warming as well as capability of power) and to verify the impact on the framework structures. This data is made open through sensors that could be put in each structure and inside the framework. If there is to be an event of region warming, building centrality profiles ought to include framework topology and characteristics that considers extreme heat to diminish the apexes. This reshaping impacts the building air temperature profile, which will be completely evaluated to maintain customer comfort at a sensible level.

Additionally, as a result of the Internet-of-Things (IoT) [7] and correspondence measures [8], watching data is made accessible at the utility organization gather points so trapped methodologies are actualized in cloud or pack structures. Another critical impact of the IoT is the clear joining of heterogeneous learning sources that will be used to make smarter methodologies.

The data which is collected from different source has been identified the accuracy level and properly fitted to the model by some meta-heuristic algorithms [9–13]. The authors has developed various home automation and energy management technique using the IoT devices [14–17].

3 Research Gap Identified

Through the audit of existing writings, certain confinements are seen, and they are characterized below:

1. In existing writings, the disadvantages or issues in innovation of a smart network alone are not assessed.
2. Research focuses on only a discrete part, which means either on network framework or power framework.
3. Methodology to defeat issues in the IoT engineering is not introduced or inquired about.
4. A useful upgrade of smart framework applications for the existing successful power board plan is not introduced.

4 Proposed Methodology

The introduced research technique deals with all the domestic appliances and related sensor data are connected through the IoT's needs, the devices are connected to local Ethernet. Then this LAN can be extended to multiple rooms or locations to monitor the energy utilization, etc. The Energy related data will be stored in a Data server on LAN, which runs a separate application called Energy Management System (EMS) to give energy advice about the domestic appliances. In this scenario, a stochastic-based energy management scheme will be developed and adopted for an effective energy saving mechanism in smart grid applications. The EMS connects through the GSM 3G network using the designed mobile application. This smart EMS can be extended over the IoT for any number of smart buildings or smart homes in the smart city for maximum comfort.

The present research is planned to be implemented in the following research phases (Fig. 1):

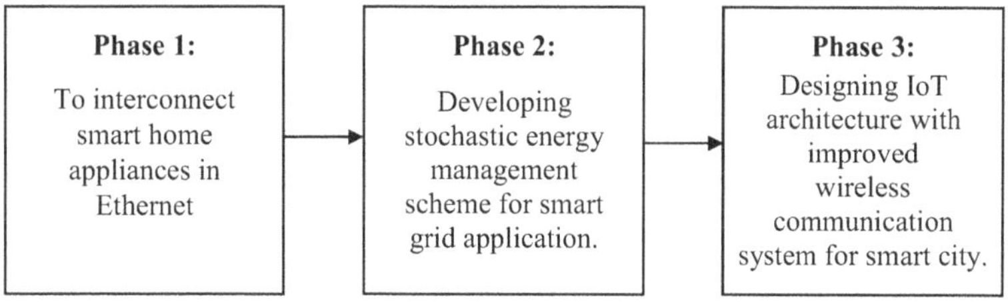

Fig. 1 Phases of research work

4.1 Stochastic Energy Management Scheme

To control the flexibilities and their uncertainties motivated a stochastic energy management scheme. To that end, the parameters $\left\{\left(p_t^c, q_t^c, \overline{p}_t^q, \pi_0, t\right)\right\}_t$ are modelled as stationary as well as processes of ergodic stochastic. The new energy management scheme is thus posed as:

$$C := \min_{\left\{p_t^g, q_t^g\right\}} \mathbb{Z}\left[\pi_{0,t}\left(p_t^{\mathrm{T}} R_{pt} + q_t^{\mathrm{T}} R_{qt} - 1^{\mathrm{T}} p_t\right) + \pi_f 1^{\mathrm{T}}\left[p_t\right]\right] \tag{1}$$

$$S \text{ to for all } t, \quad 0 \leq p_{n,t}^g \leq \overline{p}_{n,t}^p, \quad \forall\ n \in S_t \tag{2}$$

$$p_{n,t}^g = \overline{p}_{n,t}^g, \quad \forall\ n \in \overline{S}_t \tag{3}$$

$$\left|q_{n,t}^g\right| \leq \tan\ \theta_c p_{n,t}^g, \quad \forall\ n \tag{4}$$

$$\left(p_{n,t}^g, t\right)^2 + \left(q_{n,t}^g, t\right)^2 \leq \overline{S}_n^2, \quad \forall\ n, t \tag{5}$$

$$p_t^{\mathrm{T}} f_n f_n^{\mathrm{T}} p_t + q_t^{\mathrm{T}} f_n f_n^{\mathrm{T}} q_t \leq \overline{S}_n, \quad \forall\ n, t \tag{6}$$

$$v_l \leq 2Rp_t + 2Xq_t + v_0 t \leq \overline{v}_u, \quad \forall\ t \tag{7}$$

$$\mathbb{Z}\left(p_{n,t}^g, t\right)^2 + \left(q_{n,t}^g, t\right)^2 \leq \overline{S}_n^2, \quad \forall\ n \tag{8}$$

$$\mathbb{Z}\left[p_t^{\mathrm{T}} f_n f_n^{\mathrm{T}} p_t + q_t^{\mathrm{T}} f_n f_n^{\mathrm{T}} q_t\right] \leq S_n, \quad \forall\ n \tag{9}$$

$$v_l \leq \mathbb{Z}\ \left[2Rp_t + 2Xq_t + v_0 t\right] \leq v_u \tag{10}$$

Where the desires are assumed control over the joint dissemination of $\{(p_t^c, q_t^c, \overline{p}_t^q, \pi_0, t)\}_t$. The stochastic issue includes limitlessly numerous advancement factors, which will be signified $x := \{(p_t^g, q_t^g)\} t$.

The limitations in (1)–(7) apply deterministically consistently, while those in (3)–(7) compare to looser immediate operational points of confinement. Then again, requirements (8)–(10) uphold more tightly operational points of confinement, yet in a normal sense, consequently coupling factors crosswise on occasion. Observe that constraint $(p_{n,t}^g, t)^2 + (q_{n,t}^g, t)^2 \leq S_n^2$, $\forall n$ implies constraints (5) and (7), but the converse is not true.

5 Result and Discussion

The energy consumption and the cost reduction result achieved by the proposed method is discussed below with graphs and tables.

Figure 2 shows the energy consumption graphs, and Table 1 shows the values achieved by energy consumption.

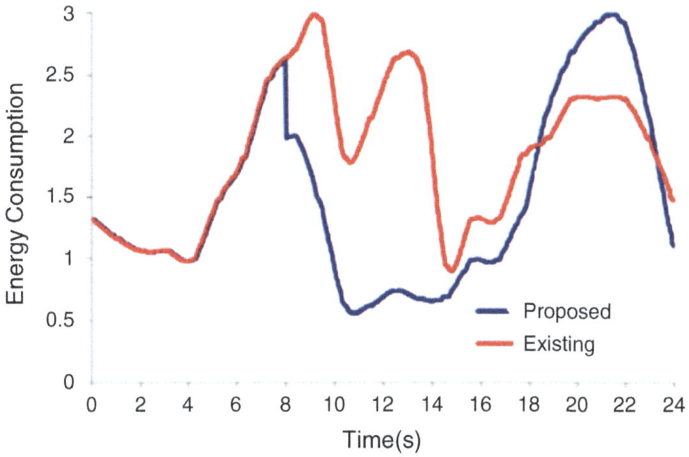

Fig. 2 Energy consumption graphs

Table 1 Energy consumption

No. of energy hub	Proposed method	Existing method
6	2.4	2.5
8	1.8	2.8
12	0.7	2.2
14	0.6	1.3
16	0.9	1.2

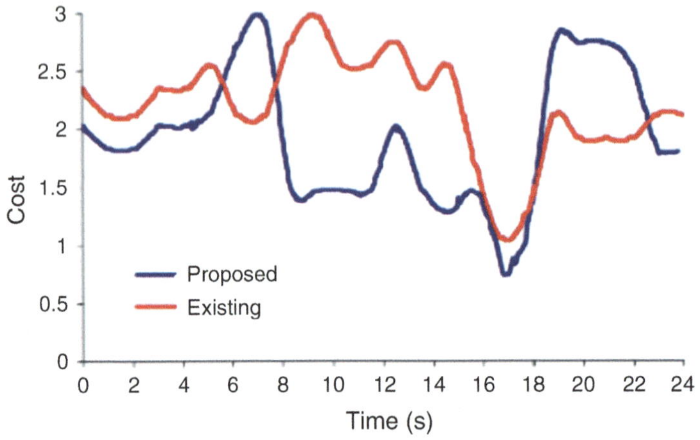

Fig. 3 Cost reduction graphs

Table 2 Cost reduction with proposed and existing methods

No. of energy hub	Proposed method	Existing method	Peak to average ratio (PAR)
6	232	245	5
8	200	235	2
12	243	276	4
14	254	285	7
16	219	227	3

Table 3 Power consumption

Hub	Proposed method	Existing method
1	62	69
2	52	67
5	48	53
7	50	70

Figure 3 and Table 2 show the cost reduction with the proposed and existing methods, and the peak to average ratio value achieved by the proposed and existing methods.

Table 3 shows the power consumption of the proposed and existing methods. Figure 4 shows the power consumption graph.

6 Conclusion

As an indispensable fundamental connection of smart brace fringe data observation, the sensor network has wide application space in the power framework, which will assume a significant job in lattice development and control framework well-being generation through the board. The intellectualization of intensity framework has

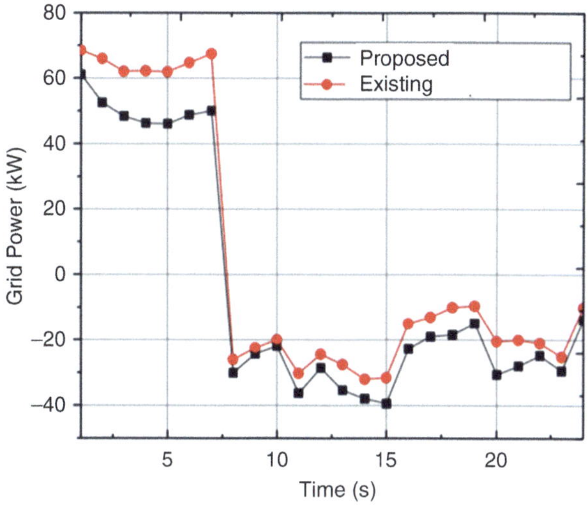

Fig. 4 Power consumption

not been finished, and a great deal of the currently used gadget should be updated because it needs a gigantic amount of work. Monstrous information handling and capacity are required for the IoT. The PC information preparing capacity must meet higher and stricter necessities, and the related equipment expenses of the intensity framework are also tremendous.

References

1. Lobaccaro, G., Carlucci, S., & Löfström, E. (2016). A review of systems and technologies for smart homes and smart grids. *Energies, 9*(5), 348.
2. Zhou, B., Li, W., Chan, K. W., Cao, Y., Kuang, Y., Liu, X., & Wang, X. (2016). Smart home energy management systems: Concept, configurations, and scheduling strategies. *Renewable and Sustainable Energy Reviews, 61*, 30–40.
3. Li, W., Logenthiran, T., Woo, W. L., Phan, V. T., & Srinivasan, D. (2016, July). Implementation of demand side management of a smart home using multi-agent system. In *2016 IEEE Congress on Evolutionary Computation (CEC)* (pp. 2028–2035). IEEE.
4. Jayanthiladevi, A., Murugan, S., & Manivel, K. (2018). Text, images, and video analytics for fog computing. In *Handbook of Research on Cloud and Fog Computing Infrastructures for Data Science*.
5. Narkhede, M. S., Chatterji, S., & Ghosh, S. (2013). Multi-agent systems (MAS) controlled smart grid—A review. *International Journal of Computer Applications*, 12–17.
6. DeFries, R. S., Rudel, T., Uriarte, M., & Hansen, M. (2010). Deforestation driven by urban population growth and agricultural trade in the twenty-first century. *Nature Geoscience, 3*(3), 178–181.
7. Ashton, K. (2009). That Internet of Things Thing. *RFID Journal*. Retrieved from https://www.rfidjournal.com/articles/view?4986

8. Giusto, D., Iera, A., Morabito, G., & Atzori, L. (Eds.). (2010). *The internet of things: 20th Tyrrhenian workshop on digital communications*. New York: Springer.
9. Sampathkumar, A., & Vivekanandan, P. (2019, August). Gene selection using PLOA method in microarray data for cancer classification. *Journal of Medical Imaging and Health Informatics, 9*(6), 1294–1300.
10. Madhavan, P., Thamizharasi, V., Ranjith Kumar, M. V., Suresh Kumar, A., Asaduzzaman Jabin, M., & Sampathkumar, A. (2019). Numerical investigation of temperature dependent water infiltrated D-shaped dual core photonic crystal fiber (D-DC-PCF) for sensing applications. In *Results in Physics (Elsevier)* (Vol. 13, p. 102289).
11. Ramana, T. V., Pandian, A., Ellammal, C., Jarin, T., Rashed, A. N. Z., & Sampathkumar, A. (2019). Numerical analysis of circularly polarized modes in coreless photonic crystal fiber. *Results in Physics (Elsevier), 13*, 102140.
12. Chen, S. L., Villaverde, J. F., Lee, H. Y., Chung, W. Y., Lin, T. L., Tseng, C. H., & Lo, K. A. (2017). A power-efficient mixed-signal smart ADC design with adaptive resolution and variable sampling rate for low-power applications. *IEEE Sensors Journal, 17*(11), 3461–3469.
13. Sampathkumar, A., & Vivekanandan, P. (2018). Gene selection using multiple queen colonies in large scale machine learning. *International Journal of Electrical Engineering, 9*(6), 97–111.
14. Soares, J., Ghazvini, M. A. F., Borges, N., & Vale, Z. (2017). A stochastic model for energy resources management considering demand response in smart grids. *Electric Power Systems Research, 143*, 599–610.
15. Yao, L., & Tsai, T. S. (2016, December). Novel hybrid scheme of solar energy forecasting for home energy management system. In *2016 IEEE International Conference on Internet of Things (iThings) and IEEE Green Computing and Communications (GreenCom) and IEEE Cyber, Physical and Social Computing (CPSCom) and IEEE Smart Data (SmartData)* (pp. 150–155). IEEE.
16. Jayanth, S., Poorvi, M. B., & Sunil, M. P. (2016, November). Raspberry Pi based energy management system. In *2016 Online International Conference on Green Engineering and Technologies (IC-GET)* (pp. 1–5). IEEE.
17. Gu, W., Wu, Z., Bo, R., Liu, W., Zhou, G., Chen, W., & Wu, Z. (2014). Modeling, planning and optimal energy management of combined cooling, heating and power microgrid: A review. *International Journal of Electrical Power & Energy Systems, 54*, 26–37.

A. Sampathkumar received his Bachelor's in Information Technology, Master's in Mainframe Technology and Ph.D. degree at Anna University Chennai. He is currently working as an Assistant Professor in the School of Computing Science and Engineering at VIT Bhopal University. He has published several articles in peer-reviewed journals and is a member of CSI societies. His research interests include Artificial Intelligence, Data Mining, Machine Learning, Data Analytic and Optimization Techniques.

S. Murugan obtained his Bachelor's degree from the Department of Computer Science and Engineering at Mailam Engineering College, Anna University. He obtained his post graduate degree from the Department of Computer Science and Engineering at Hindustan University, Chennai. Currently, he is pursuing his Ph.D at Mewar University, Rajasthan. His specializations include Wireless Sensor Networks, Network Security, Cryptography, Database Management Systems and Data Mining. He has published papers in SCi Indexed, Scopus Indexed and other International Journals and Conferences.

M. Sivaram completed his B.E (CSE) at Bharat Niketan Engineering College, Madurai Kamaraj University, Madurai in 2002. He was awarded M.Tech (CSE) degree from the National Institute of Technology, Trichy in 2007. He completed a Ph.D degree in Information and Communication Engineering from Anna University, Chennai in 2014. He has nearly 18 years of experience teaching in both the UG and PG program. He is presently working as a Professor in the Department of Information Technology in Lebanese French University, Erbil. His fields of interest are Data Mining, Image Retrieval, Information Retrieval, Data fusion, Image Processing and Artificial Intelligence. He has published more than 30 papers in International Journals and Conferences.

Vishnu Sharma completed his B.Tech, M.Tech and Ph.D (CSE) from Madhav Institute of Technology & Science (M.I.T.S.) Gwalior (M.P.) in Computer Science & Engineering and is affiliated with Rajiv Gandhi Technical University, Bhopal. He is currently working as a professor/head in the Department of CSE at Galgotias college of Engineering and Technology. He has published more than 53 research papers in the area of Mobile Adhoc Networks and Mobile Computing in International/National Conferences and International/National Journals. He has also published three books on Mobile Computing, Advanced Mobile Computing and Cyber Security. He has around 21 years of teaching experience in various reputed engineering institutes and Universities. He has been involved in organizing several international/national conferences and workshops. He was the editor of IEEE Conference ICCCA proceedings. He was the Conference Technical Chair of several IEEE conferences.

K. Venkatachalam received his Bachelor's in Information Technology in 2005, Master's in Computer Science and Engineering in 2008 and PhD in Computer Science and Engineering in 2018. He has more than 12 years of academic experience and currently works as an Assistant Professor (SR) in the School of Computer Science and Engineering at VIT Bhopal University, Bhopal, Madhya Pradesh, India. He has published several articles in peer-reviewed journals and his research interests include Data Mining, Web Services, Semantic Web Services, Distributed Computing and Cloud Computing.

Manikandan Kalimuthu completed his B.E (CSE) from Madras University in 2002, M.E (CSE) from Annamalai University in 2006 and Ph.D in the field of Image Processing from Anna University in 2016. He has about 17 years of teaching experience at various levels and presently works as an Associate Professor in the School of SCSE, VIT Bhopal University, Bhopal. He has published 30 papers in International Journals and International Conferences. His research interests include Image Retrieval, Face Recognition, Pattern Recognition and Information Security.

Smart Vehicular Management System Using Blockchain for Internet of Connected Vehicles

K. S. Umadevi, B. Shalu Mohammed, P. Balakrishanan, K. R. Jothi, M. Madiajagan, D. Geraldine Bessie Amali, and P. Madhavan

1 Introduction

The Internet of Things is the next emerging technology, where many devices could be interconnected and interact with the environment to collect data and information. The Internet of Things (IoT) [1] requires a complete and rich set of wireless connectivity and networking protocols. The Internet of Things is from wearables devices to smart appliances to automobiles, where connected devices also require scalability for resources with constrained devices. All embedded software systems and services use a rich set of connected devices for the IoT that meets certain conditions, such as reliability, security and power consumption for demanding markets, including medical, industrial, automotive, etc. All the IoT solutions are enabling the development and control of a connected embedded system very safely and securely.

Blockchain is a decentralized computation and information sharing platform that enables multiple authoritative domains, which do not trust each other, to cooperate, coordinate and collaborate in a rational decision making process. Blockchain was first introduced in 2009 by Nakamoto [2]. Blockchain has been developed over many years for usage other than just cryptocurrencies. Blockchain is the first historic permanent decentralised, global trustless ledger of records. Since the invention of blockchain, all the entrepreneurs of different industries around the world have come to know about the implication of blockchain development. The idea of blockchain technology has its own imagination running wild, and it can be applied to anywhere

K. S. Umadevi (✉) · B. Shalu Mohammed · P. Balakrishanan · K. R. Jothi · M. Madiajagan
D. Geraldine Bessie Amali · P. Madhavan
School of Computer Science and Engineering, Vellore Institute of Technology, Vellore, India

Department of CSE, SRM Institute of Science and Technology, Kattankulathur, Tamil Nadu, India
e-mail: umadeviks@vit.ac.in; madhavap@srmist.edu.in

G. R. Kanagachidambaresan et al. (eds.), *Internet of Things in Smart Technologies for Sustainable Urban Development*, EAI/Springer Innovations in Communication and Computing, https://doi.org/10.1007/978-3-030-34328-6_13

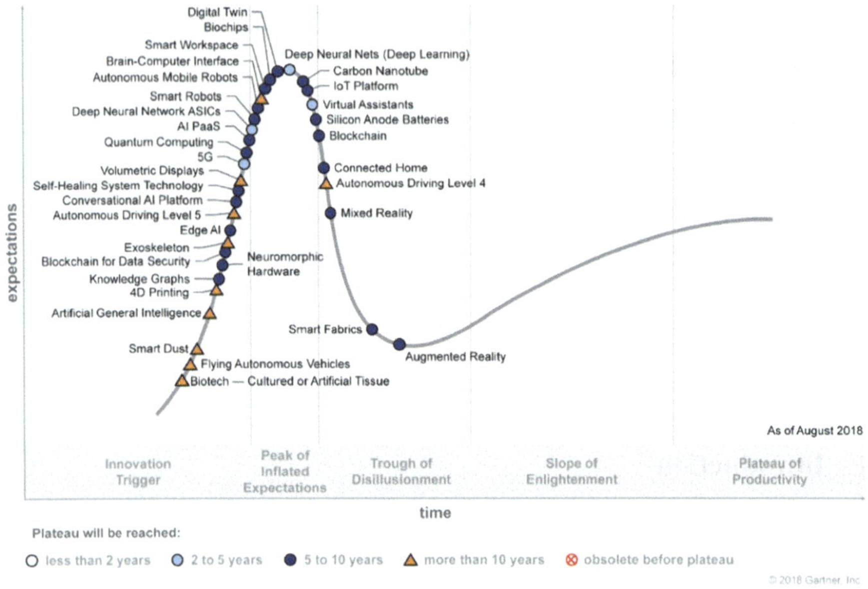

Fig. 1 Gartner hyper cycle [3]

that has need of trustworthy records. There is complete power for cryptography for every individual work and to stop the third party digital relationship by requiring authority of transactions considered as ‘pull transactions’ (Fig. 1).

The concept of the Internet of Things (IoT) refers to connecting products/processes that are capable of communicating through an internetwork. It follows the rule of thumb “anything that can communicate can be connected” for sharing information, i.e. gathered or processed by itself. The leading research and analytic firm Gartner [3] predicted that nearly 20 billion IoT devices will be connected by 2020. For extensive connectivity, the IoT makes use of the Internet Protocol (IP) widely used by researchers, academics, industrialists and others for internetworking. The main objective behind the establishment of the IoT is to process any data gathered by these sensors/actuators at the right time to make it meaningful with less intervention by human beings. Applications of the Internet of Things include transport and logistics, business, communication and entertainment media, health care, habitat monitoring sector, etc. Owing to the growing interest of people in benefitting from this network, industrial personnel have established a smart connectivity among the devices with the complete help of advanced networking technology such as 5G [4]. Not only the backhaul network used for facilitating the network but also the architecture and interaction pattern of the software used and the tools used for analytics aim towards smart behaviour.

IoT devices make use of the Radio Frequency Identification (RFID) microchips that enable wireless communication. RFID are most commonly used in the tracking industry without line of sight communication. The RFID system makes use of a reader to detect the tags in the objects under close proximity. The passive tags are not battery operated, but instead they make use of the readers interrogation signal to communicate the ID to the RFID reader. One of the major applications of RFID is to identify vehicles in the Intelligent Transportation System (ITS). The proposed work makes use of RFID for vehicle identification, traffic monitoring and parking. Blockchain is a sequential data structure consisting of blocks that contain multiple hashed transaction values. Any number of blocks can be added in the chain and end by using some mining techniques, whereas modification in the mid would require subsequent changes. These are duplicated across the network nodes and shared via Peer-to-Peer file transfer protocol. Though the interest behind using blockchain was initially bitcoin and the other virtual currencies, it attracted industry personnel for a variety of reasons. Since it is distributed in nature and serves as an open ledger, it is applied not only for financial applications but also business, healthcare, registrations, voting, music media, etc. The new beginning of the integration of blockchain and the IoT would be one of the most viable solutions to most daily challenges and problems. One such problem in metro cities is handling increased vehicle traffic, which results in the need to develop a Smart Vehicular Management System. Although a number of projects have been carried out for developing systems to ease the various mechanisms and stages in the vehicular field, most of them seem to be monetization centric.

This work proposes a Smart Vehicular System that is more of an application for ease and traffic reduction than a profitable business solution. According to a special report by skyline parking, on average a vehicle wastes about 2 min waiting in the parking line. About 300 vehicles go through the same process each day at one parking site. This leads to a total wastage of 600 min and excess fuel consumption. If 10 h of time is consumed per parking lot, then the amount of time and fuel wastage is a huge number across the country. In addition, around 30% of vehicles are roaming around the city to find a parking space. Hence this work is an attempt to reduce this drawback by automating the process of parking, thereby avoiding the waiting time of the vehicle using the concepts of blockchain in the IoT (Fig. 2).

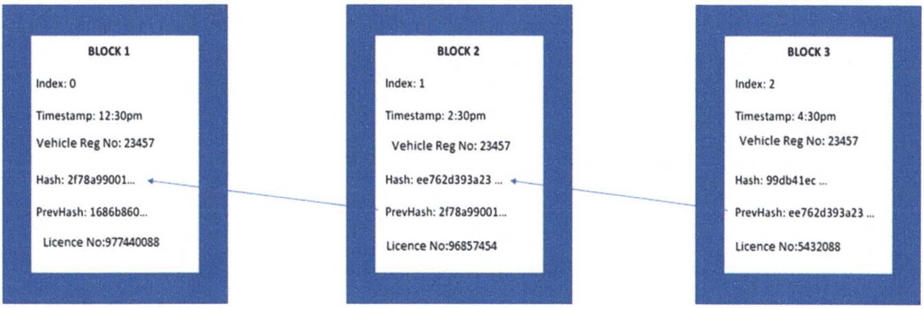

Fig. 2 Example for blockchain

The objective of this chapter is

- To develop a system for automatic parking fee transactions in various places, such as malls, hospitals and any public parking space and other buildings.
- To enhance the insurance monitoring system, to intimate the user beforehand with the deadlines and automatic pre-booking of insurance documents before expiry.
- To provide a Secured Storage of documents in the blockchain database for easy retrieval of e-documents and eliminating the need for hardcopy.
- The advantage of the system is to reduce the traffic congestion and the user can check before coming to any parking lots.
- The system will have a wallet so that every user needs some minimum balance in their account before reaching a parking space.
- Once the user leaves the parking lot the amount will automatically be calculated and reduced from the wallet, so the user does not have to wait for payment manually.
- The key concept behind using the IoT technology is RFID. Each vehicle is identified by a unique eight digit Hex-decimal number.
- The proposed system provides real time availability of parking, booking and reservation.
- The system will be able to provide advance payment using blockchain transaction.
- The city traffic congestion and other details will be updated in the system

As mentioned earlier, there are many systems formed to avoid traffic congestion in malls, hospitals, public parking spaces, etc. Some statistics have been collected in India that show the importance of reducing environmental pollution, paper waste, etc.

As seen in Tables 1 and 2, most of the issues are related to different physical and environmental problems, and they become larger when people ignore them, which in turn leads to health and environmental degradation and loss of capital.

Many environmental and economic disadvantages have occurred because of traffic congestion and traffic jams, which we seldom realize. Another major problem is that a majority of people are wasting fuel looking for parking spaces, and sometimes parking spaces are found in crowed places such as markets, leading to accidents and further frustrating drivers. Also, when drivers are looking for a

Table 1 Economic disadvantages to drivers

Events	Time (h)
Average loss of time in finding proper parking spot in major cities, such as Delhi, Mumbai, Kolkatta, Chennai and Bangalore.	44
Average loss of time in finding proper parking places in Kochin, Thiruvanathapuram, Madurai and Tirupati etc.	17
Traffic congestions in metropolitan cities	46
Traffic congestions in normal cities	15

Table 2 Non-economic and health disadvantages

Events	Percentage (%)
Arguments and fights with other people or fellow drivers and causing traffic issues	65
Stress and anxiety of female drivers	69
Missing important meetings, appointments, trains, flights, etc.	40

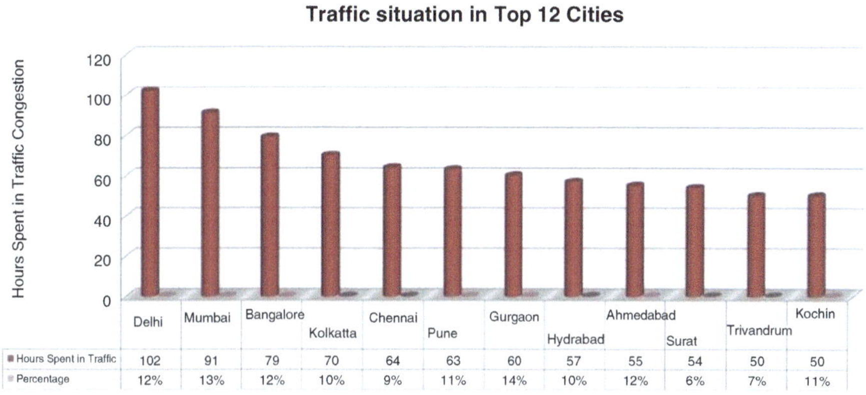

Fig. 3 Index with hours spent by drivers

parking lot they drive using lower gears, causing more emission of carbon dioxide gas into the atmosphere. Emission of carbon dioxide causes global warming and other problems to ecosystems.

The Ministry of Road Transport and Highways 2017–2018 survey [5] provides a clear view about the traffic congestion in top cities, which are almost all currently facing the same issues. Figure 3 shows hours spent by drivers.

People are spending more time looking for parking spaces and end up with no space. Parking areas nowadays have become a rare commodity, especially in metropolitan cities at certain places, such as hospitals, malls, theatres, market places, etc., and even when a space is found one must pay for it. Currently, finding a parking space at a desired location is a matter of luck and a very infrequent occurrence. Many researchers are working on blockchain technology for different areas, and a significant amount of work has already been done. Shardoor proposed the Reservation-based Smart Parking System (RSPS) [6], which introduces the idea to permit all drivers to effectively find a location and reserve vacant spaces. The system used a cluster algorithm that can help to study periodical parking status from a given sensor network, which is already deployed in parking areas, and sets up a reservation service that drivers are allowed to access with their personal communication devices. This software was implemented with a very cost efficient parking vehicles system for different levels of a parking facility and made available using WSN techniques. Also, this software was developed for a mobile android application and a cluster was used for allocation method and to perform all automatic billing processes. This system will monitor the availability of idle parking areas and guide the vehicles to the nearest free areas. Without sacrificing the reliability, they have reduced the number of sensors, which will also minimize the cost.

Dr. Anthony Mwebaze proposed the Intelligent Parking System [7] based on an integrated architecture where:

1. Wireless Sensor networks (WSNs) using Small Programmable Object Technology (SPOT) were launched into parking areas to monitor all the activity happening in parking spaces through light intensity sensing and
2. It collected all the sensed information, which were gathered and channelled through a particular gateway into a server. Any particular database could be used for information and to send it back as a parking space visualization over World Wide Web technology, and it could also connect with mobile devices via Multi-Protocol label Switching (MPLS) networks. Using a given simulation model of this parking system, which has built around a new generation of SUNspot motes, there are some studies demonstrating how the smart parking system implements in real life.

Chandnipatel described the Rotary Automated Car Parking System [8] that explains an idea about changing and reducing the working model of a vehicle parking system for parking 6–24 vehicles within a 3.17 m parking area. It is an amalgamation of existing and developing parking systems with added advantages to reduce spaces occupancy by the design of a simpler and more compact parking space that allows for rotary and vertical parking spaces.

Here, the system uses a chain and sprocket mechanism for driving to the parking platform, and around a one fourth hp brake motor can be implemented for powering the system and indexing the parking platform. The platform can be fabricated to set the working model. All the manufactured and procured items are ready to be delivered and assembled.

Ramneet Kaur and Balwinder Singh purposed the Field Programmable Gate Array (FPGA) [9] vehicle parking system. The parking area is adjusted with Liquid Crystal Display (LCD) monitors, and when a vehicle enters the parking area the available space is shown and then the gate for the vehicle entry will open. An RF module is also used to get and receive parking information. All the host computers can act as complete control units. The host computer will activate with parking slot checking and identifications. There are two modules in the system, i.e. the identification module and the parking slot module. Identification modules identify all the users who appear. Parking slot modules check parking availability status. These modules are modelled in Hardware Description Language (HDL) and formed into one unit called FPGA. This parking system was designed with different prototypes with different interfaces, such as sensor interfacing, stepper motors and LCD. Zishan Raza developed a web application about this system called "Park Easy", which is also available for mobile phones. Another advantage is that sensor monitoring techniques are implemented by a camera that also uses a sensor to take photographs and to show all the parking spaces for vehicles.

Another system was developed by P. Dharma Reddy, A. Rajeshwar Rao and Dr. Syed Musthak Ahmed using Artificial Intelligence (AI) [10] to collect images to identify vehicle occupied parking spaces. The idea of the system provides guidance for an array of images to assign specific slots, thus using some intelligence. They used the RFID method to identify the car parked in each slot, and they also provided a variable display screen on the base floor. The system reserves and assigns an optimal parking space based on cost function and parking cost. They also provided a Driver request processing center, which has the entire infrastructure for a vehicle to communicate and assign parking slots.

2 Architecture of Proposed Systems

With the multiplying advances in various fields of technology, changes in logistics, academics and automobiles are inevitable. However, these changes do not always function as smoothly as expected. Some drawbacks are:

- The rapid growth in the automobile industry and the infinite number of vehicles per square foot replacing human beings.
- Decreasing traffic space and increasing traffic congestion.

The principal issues in most of the metro cities due to the increased number of vehicles include traffic jams in malls, public roads and parking lots. In this ongoing situation, a technological automation to monitor and maintain the traffic is the need of the hour. Smart Vehicular Systems prevail at improving the efficiency of vehicular systems; however, the aftereffects of an efficient system are often negated.

Many systems are developed for controlling traffic and providing proper parking space. Researchers have come up with different solutions, but somewhere something has been missing so that no system is working properly in any metropolitan city. Vehicle management systems face many problems in cities. With the increase of population and the number of vehicles, it will be more difficult to accommodate all the vehicles on roads and to control traffic not only on roads but also in other places, such as malls, hospitals, markets, streets, etc. Now, every city needs to have parking spaces with proper parking fee management. Parking vehicles and transferring money should not be delayed; it should be fast and reliable.

The proposed work aims towards developing a system that identifies a vehicle based on its unique ID using RFID readers. It makes an attempt to integrate the big hub technologies Blockchain and the Internet of Things for developing a viable Smart Vehicular System. The Internet of Things concept is used to maintain a log of the vehicles identified in various parking lots along with the time duration in the corresponding parking space. Blockchain is used to make automatic and secure payments for parking based on the unique IDs of the vehicle and that of the parking area. The gateway to the decentralised system is the Ethereum wallet, which is deployed in Ethereum Blockchain. The basic idea is to hold and secure each ether or other crypto currency, which will be built on Ethereum, and to deploy it with the help of Smart contracts.

The main advantage of the proposed work is digital transaction using blockchain. Another advantage is to remove third party involvement for parking payment transaction. The major befits are:

- Transactions will be done within a fraction of a second so a driver does not have to wait.
- Transaction will be done on driver's phone through wallet transaction
- No need to wait for any queues for payment transaction
- There is no hassle of cash
- E-receipt will be generated

2.1 Smart Vehicular System

A Smart Vehicular System is an application to ease parking and traffic reduction rather than a profitable business solution. According to a special report by skyline parking [11], on average a vehicle wastes about 2 min waiting in a parking line. About 300 vehicles go through the same process each day at one parking site. This leads to a total wastage of 600 min and excess fuel consumption. If 10 h of time is consumed per parking lot, then the amount of time and fuel wastage is a huge number across the country. In addition, around 30% of cars roam around the city looking for parking spaces to avoid traffic congestion. Hence, this work is an attempt to reduce this drawback by automating the process of parking, thereby avoiding the waiting time of the vehicle using the concept called Blockchain.

With the multiplying advances in various fields of technology, changes in logistics, academics and automobiles are inevitable. These changes do not always function as smoothly as expected. One such drawback is the rapid growth in the automobile industry and the infinite number of vehicles. The increased number of vehicles proves to be one of the principal issues in most of the metro cities leading to traffic jams in malls, public roads and parking lots. In this prevailing situation, a technological automation to monitor and maintain the traffic is the need of the hour. Smart Vehicular Systems prevail at improving the efficiency of vehicular systems; however, the aftereffects of an efficient system are often negated. The proposed work aims towards developing a system that identifies a vehicle based on its unique ID using RFID readers. It makes an attempt to integrate the big hub technologies Blockchain and the Internet of Things for developing a viable Smart Vehicular System. The Internet of Things concept is used to maintain a log of the vehicles identified in various parking lots along with the time duration in the corresponding parking space. Blockchain is used to make automatic and secure payments towards parking based on the unique IDs of the vehicle and that of the parking area. The gateway to the decentralised system is the Ethereum wallet, which is deployed in Ethereum Blockchain. The basic idea is to hold and secure each ether or other

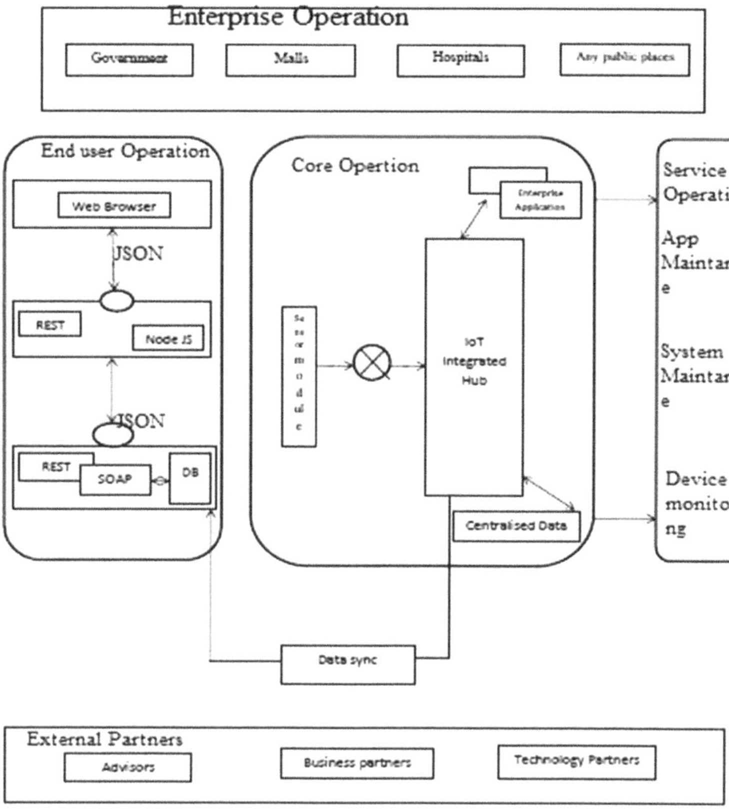

Fig. 4 Architecture of Smart Vehicular Management System

crypto currency that will be built on Ethereum and to deploy it with the help of Smart Contracts.

The designed architecture of the Smart Vehicular Parking System is illustrated in Fig. 4; there are five major parts:

1. The sensing network
2. Blockchain with hash functions
3. The user interface for data management and visualization
4. Payment gateway
5. Have been reported for completeness, even if its development is obviously outside of our work.

The sensing network contains the RFID-WSN 6LoWPA network, which is composed of three types of nodes: (1) 6LowPAN Border Router (6LBR), (2) 6LowPAN Router (6LR) and (3) 6LowPAN Router Reader (6LRR). Based on 6LoWPAN standard, 6LBR is mainly used to connect the entire system to the Internet by changing 6LowPAN packets to IPv6 packets and vice-versa. A 6LR gives details about a node that is able to provide forwarding and routing capacity, and 6LRR is defined as a 6LR interfaced with the connection of RFID reader. The system assumes that each 6LR is equipped with a light sensor that is placed on every parking area to check for vehicle presence and 6LRR nodes are placed on every pole to locate the nearest parking area that is reserved for people with special payment amounts and authorizations. Indeed, in the proposed system, the 6LRR nodes are used to check only for authorized vehicles.

Once the information is collected then the retrieval of information can be delivered to the IoT Smart Gateway, which is already connected to the sensing network and, also, with the Internet through a 3G-communication interface. The communication gateway is a role of 6LBR, which enables all communication between Hybrid Sensing Network (HSN) nodes and all end users. Here, initialization of the network will be the execution of a smart contract and choosing proper tools. The idea of a smart contract is to implement blockchain. There would be already created blocks and other blocks would be added and chained with all the characteristics. After all the blocks are chained they will start sharing through networks. All the blocks are created using Ethereum and different parameters are specified: "nonce" (generally cryptography hashes generates a random values), "timestamp" (validation of two blocks and their timing). Once Ethereum is formed then every user is responsible for creating folders containing blocks and initializing in it. The purpose is to replicate all commands in the network with n number of times that are already connected with nodes and an agreement being set. The communication between two blocks can be avoided with respect to their agreement using a smart contract. Secure hash algorithm (SHA-2) is a cryptographic function that is designed and published by the United States National Security Agency (NSA) [12].

The Merkle–Damgard [13] structure was used to build the system from the one way compression method itself, and the Davies–Meyer structure was used for specialized block text cipher. Cryptographic hash functions have a pure mathematical operation, which has runs on digital data; by comparing runs the computed hash functions called "hash" (output from execution of algorithm), which will be known, and the expected hash values can determine the data integrity. For example, comparing two hash functions, one hash function is downloaded and the other one is a previously published hash and the result can show whether the downloaded hash has been modified or tampered with. A Public Ledger can be defined as that originated by old records used to store information, such as agriculture prices, news and all other analysis.

A Public Ledger is based on the availability for general public viewing and for verification. Generally, a public ledger is defined as data management or some server stored system just like a database system of any particular bank's records. Here, blockchain is also a form of public ledger, which has a series of blocks that contains all the information transaction details that are recorded with suitable authentication and verification for the designed network participants. After all the information is collected, it will be stored as records and transactions in a public ledger, then the exact working of cryptocurrency will start. When each block is filled with transaction details and new blocks could be mined then that will be added to the next block by network participants called miners. Here, thousands of such participants could maintain the copy of the public ledger so that everyone has to get to know the true status of the entire network and how many people hold crypto tokens also; all transactions must be authentic to be recorded and to prevent any misuse such as double spending. There are many combinations regarding public ledgers based on their algorithms, such as consensus algorithms, encryption and reward mechanism. These are all used to identify the network participants, which are protected, and all genuine and validated transactions are carried out on the network for further processes.

The customer operations and front office operations will be made available on all popular devices like iPhone, Android, Windows and over the Web. Data will be served over a secured cloud network. The core component consists of IoT sensor modules, which relays events to the IoT gateway. The IoT gateway aggregates the event and transforms the data into a machine-readable format. With the help of machine learning and data analytics, the system will be able to optimize the parking search over time and provide the most accurate locations in the least amount of time on customer's devices. Back office operation will be available to support customers 24/7 with any issues or concerns they have regarding their bookings, use of the system, use of consoles, or any other technical difficulties. The core part of this architecture is blockchain, where the trust, immutability and anonymity come together and are built by default. One of the responsibilities is to accept the transaction on the network. The blockchain protocol provides a maximum fault tolerance and the driver or user needs to advance GAS payment in the contract. If transaction is done and it updates to block then the miner will confirm about the transaction.

Smart vehicular system provides two types of users.

1. User or Drivers
2. Owners

Users/Drivers Users are guided through application or navigate to the proper parking lot if they booked before coming to the parking space. Every user is provided with an RFID, which will scan once the vehicle enters the parking space. Once RFID scans everything and is verified, then the user will get access to enter the parking slot. Once the user enters the system, the timer will be triggered and calculate the in-time. Once the user leaves the parking slot, the system will trigger the out-time and calculate the cost and return the amount. Once the user shows their RFID then the transaction will be carried out and the amount will be detected from the wallet. Here, the transaction is carried out through Smart Contracts, so the user will never have to worry about queues and traffic at any place.

Owners The Owner is the one who has all the control over smart contract and blockchain transactions. The user wallet will be integrated with the owners blockchain. Owners are hospitals, malls, any public parking, etc. All the owners are connected through blockchain so it would be able to manage and spot the parking

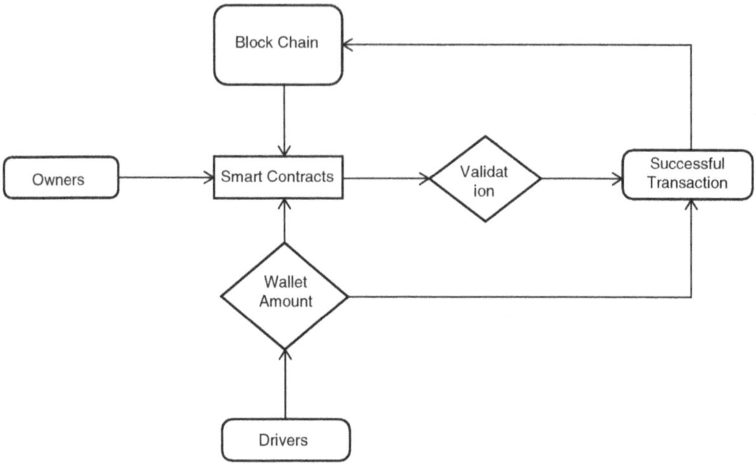

Fig. 5 Architecture of smart contract

space directly. More visibility on the global platform is a medium to monetize every unused parking space, and every owner can increase the revenue by reducing traffic congestion and redundancies (Fig. 5).

3 System Design

The IoT and Blockchain are integrated into the Smart Vehicular System to build an intelligent system for monitoring vehicular parking and insurance systems. The aim of the proposed work is to utilize the technologies to the best of their abilities in building a system useful in day-to-day activities. The newly arriving vehicles must register themselves through the manager application with the following functionalities:

- Vehicle Information Node: The vehicle registration number is used to create an Identity by the Smart Vehicular Management System and updated in the ledger using a smart contract.
- Parking Payment Node: Using entry and exit, the payment is calculated.
- Payment Node: Two payment functionalities are monitored, namely parking charges and insurance payment

Smart Contracts: Traditional contracts are still used to date with written text or file documents that framework cooperation, which has some terms and conditions and privacy policies with partnership agreements. Basically, traditional contracts are written in human language or any regional language. Here, the third party enforces the law that can be decided during some dispute for instance and it is the normal way that the third party is involved in all smart contracts (Fig. 6).

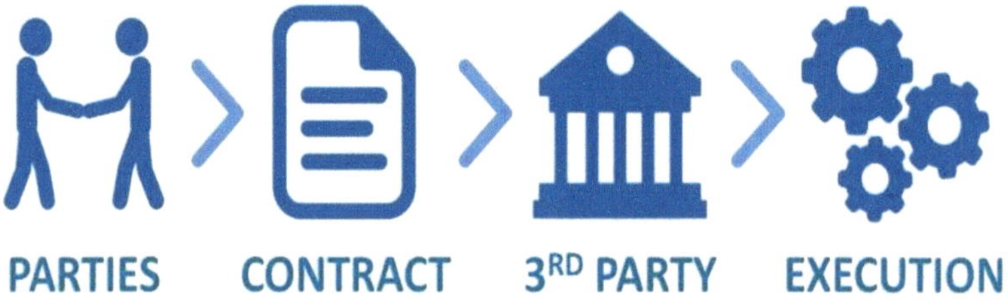

Fig. 6 Traditional contracts

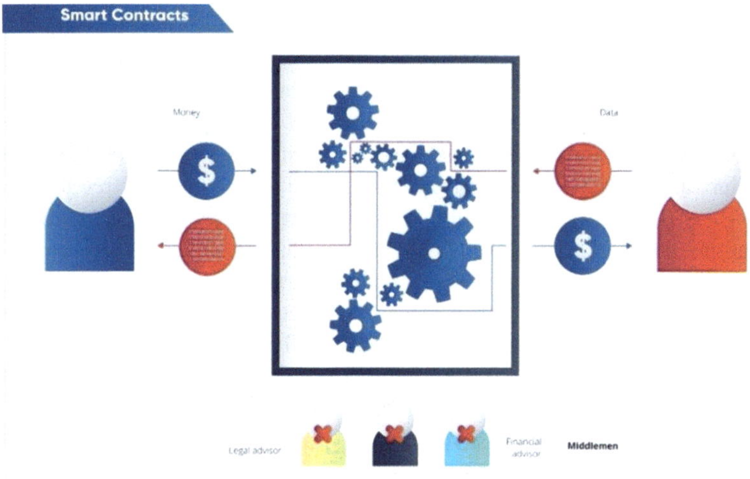

Fig. 7 Smart contract without middleman

The major disadvantages are:

1. Time consuming
2. Resources consuming
3. Costly

Smart contracts deal with revoking payment information based on deadlines of insurance. Parking payment is calculated using the entry and exit time of the vehicle and suitable communication is made to the user. A smart contract is basically a written set of programming codes that will be embedded on the blockchain. A smart contract code consists of all kind of rules, conditions and history, expiry date and all relevant information needed and is executed automatically when all the terms and conditions are met. The concept of a Smart Contract is compared to digital vending machines (Fig. 7).

Smart contracts are agreed upon and stored on a decentralized encrypted ledger. In the decentralized blockchain or distributed ledger technology, anyone can create their own database where multiple and mutually untrusting users can exchange transactions or append records without any third party coordinators.

3.1 Application of Blockchain

One of the essential features of a smart contract is its relation to users and how it can affect all the aspects of a business for customer industries. The basic idea of a smart contract is for a particular business process to be deployed to the blockchain to perform sophisticated functions. Usually, blockchain applications are referred to as Decentralised Application ('dApp'). The context of blockchain applications can be understood as blockchain-based user interfaces that can be connected to end users technology and that have the combination of smart contracts.

Smart contracts can help all participants to exchange money, property or anything they value, and it will be monetarised and verified. It is a conflict free method that cuts out any central entity and removes third party involvement. Traditional web applications and blockchain are somewhat similar, where traditional use HTML, CSS and JavaScript, and blockchain uses the smart contracts to render a particular page. Decentralized applications use smart contracts in order to communicate to the blockchain. Decentralized applications are basically runs on Peer-to-Peer networks rather than a single system. The concept of a decentralized application occurred when P2P network existed. It is a software program that will not have any single entity control. dApp are always runs on top of the blockchain networks. A smart contract consists only in the back-end and is small. If there are any chances to create dApp then different smart contracts need to be created that rely on a third party system for the front-end side (Fig. 8).

It provides a decentralised public digital ledger of transactions, which tracks ownership securely in an environment without trust. While the first applications of blockchain technology were for digital currency transactions, that same technology is now finding application in the development of the dApp, including recording the movement of data, registering unique user names and even data storage. There are also cryptocurrencies themselves being deployed to help create the dApp.

4 Implementation

To develop the proposed system, a number of frameworks are used to fit in the various processes and assemble them. The ADEPT () project is used as the foundations of the proposed work.

- Arduino: It is an open source electronic platform that has easy to use hardware and software. Arduino boards are used to get electronics inputs from users, and get the output as information. It is basically used to send a set of instructions

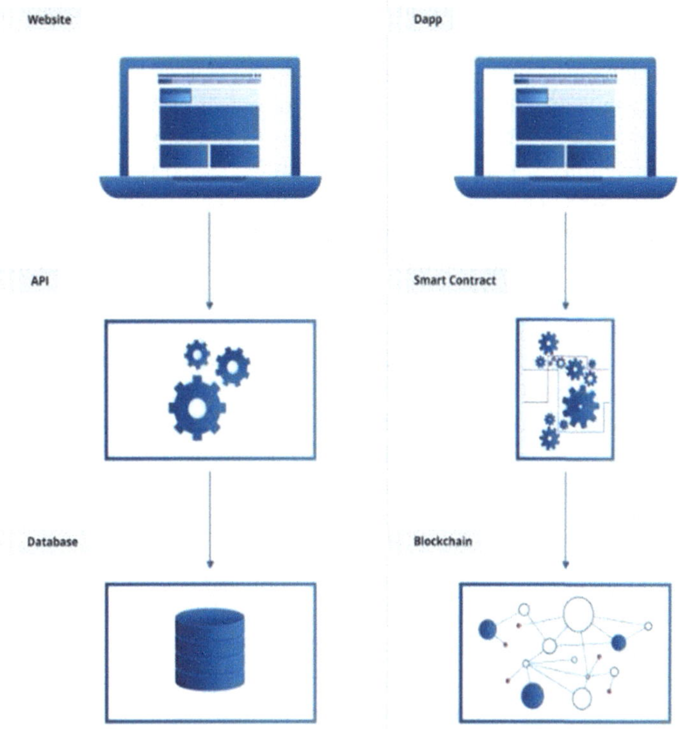

Fig. 8 Traditional web vs. decentralized web

from the microcontroller to the board. Basically, Arduino programming uses embedded C. The advantages of using Arduino are:

- Inexpensive
- Simple and clear programming environment
- Open source and extensible in both software and hardware platform
- Cross-platform

• SHA256 package: SHA256 is one of the powerful hashing mechanisms used in security. Hashing is the method of converting a data into a Finite set of hash characters.
• RFID Module: RFID or Radio Frequency Identification is a small chip that is used to unlock some keys. This chip is integrated with an RFID reader to perform some system task. RFID data is stored electronically and data retrieval uses electromagnetic waves. This RFID can store some kilobytes of data as well. RFID has the ability to perform Read, Write and Update operations in a single task.

- IR Sensor: Infrared Sensor is an electronic device. Sensors are basically used for sensing data from surroundings. IR sensors are able to measure the heat and motion of the object and to detect the motion and heat using infrared radiations.
- Solidity Language: Solidity language is pure object oriented programming and is used to create smart contracts. It is used to implement various blockchain platforms. Solidity was developed by Gavin Wood and many former Ethereum core contributors. The language is only executable in bytecode and its works through Earned Value Management (EMV). One of the advantages is developers can write their own smart contract and implement self-enforcing business logics. One who has basic knowledge in programming language can easily write a smart contract using solidity language.
- Atom Framework: It is an open source Ethereum Platform, and it is developed and maintaining by Github. Atom works to solve real time problems and for knowledge sharing for good productive quality software. Node JS is used to run back-end work, and it also helps for front-end work with web applications. One of the important advantages of using atom framework is the web browser and components that are connecting have inter-process communication among themselves.
- MetaMask: MetaMask is the Browser extension and is used for Ethereum wallet for money transaction. It is a DApp or Decentralised Application on the Ethereum network and is used as the user gateway for money transferring.
- Rinkeby Faucet Network: This is used for fund transferring in ethers. Basically this site is used for preventing any kind of malicious attack or DoS attack from the external sources. The person should need a Facebook or Google plus or Twitter account for request funds and it would permit to transfer the ethers to MetaMask wallet.
- Infura Network: It is a tool provided by Ethereum so anyone can create their own application and connect to the Ethereum blockchain. Each user created node will always connect and interact with the Ethereum blockchain. It is one of the Etherum applications that is easy for developers to create and interact with blockchain. The three major advantages are fast access, scalable and data storage.

5 Methodology

The IoT and Blockchain integrated Smart Vehicular System revolves around the integration of the two technologies to build an intelligent system for monitoring vehicular parking and insurance systems [14, 15]. The aim of this work is to utilize the technologies to the best of their abilities in building a system useful in day-to-day activities. The newly arriving vehicles must register themselves through the manager application. Then the following functionalities will take place.

- Vehicle Information Node: Vehicle registration number is used to create an Identity by the Smart Vehicular Management System and is updated in the ledger using smart contract.
- Insurance Node: Accesses the information about insurance policy of the vehicle and alerts the registered number about payment and due dates.
- Parking Payment Node: Using entry and exit, the payment is calculated.
- Payment Node: Two payment functionalities are monitored, namely parking charges and insurance payment

Smart Contract: This module deals with revoking payment information based on deadlines of insurance. Parking payment is calculated using the entry and exit time of the vehicle and suitable communication is made to the user. The key modules used in the work include:

Identification of vehicles: The key concept behind the use of the IoT technology is called RFID. Each vehicle is identified by a unique 8 digit Hex-decimal number of the form AB: CD: EF: GH. On detection of a vehicle, the vehicle is mapped with the location's unique RFID. The time stamps at the time of arrival and departure are calculated. This mechanism is used for detection and identification of vehicles based on their unique ID and passing the ID to the blockchain. This provides a low range detection mechanism, which is a viable solution in parking lots. Since the project is a large scale deployment, the monetary system must be efficient. RFID is one such method and gives a cheap solution. RFID is passive and does not involve the need for batteries. This not only helps in faster detection but also saves power consumption. The RFID chip is installed in every parking place. An RFID reader is located at the parking locations. The reader scans the chip to identify the unique ID. The unique ID identified is used as the key value in the blockchain. When a new ID is generated a new block is created. Upon identification of an already existing block, the time stamps are recorded. The unique IDs of the vehicles and the locations are used as two separate blocks in the chain and transactions take place between the blocks, which are in turn updated on all the systems by means of a hash value.

Secured Payment at parking lots, malls and hospitals: The underlying concept for secured transaction involves blockchain for monetary exchange. The RFIDs of the vehicles and the parking locations are mapped one on one. This facilitates a pre-determined amount transfer based on the parking charges that are based on time limits. A new hash value is computed each time a transaction occurs or a change occurs, thus maintaining integrity of information. Since the transactions are purely based on the ID and facilitated through blockchain, the transactions are secure. Blockchain contains each vehicle with its unique ID as an individual chain and the various chains are interconnected. Blockchain is created using node.js. Each block consists of a unique ID, which could be associated with either the vehicle or a parking location. The vehicle arrival and departure time stamps are also computed by the system clock and stored in a variable. The difference between the time stamps are used to calculate the duration and the amount is deducted from the wallet and the transaction is made. When a transaction occurs between two individual blocks, the transaction is added to all the blocks within the chain and the hash value is recalculated to maintain security. The recomputed hash value is unique each time, thus helping maintain integrity and privacy.

Secured saving of documents online for easy and timely access: The underlying architecture involves the use of a cloud based database. The public ledger is used for saving the documents in a decentralized fashion. It is also scalable and provides real-time access to stored data, which is a must for the number of vehicles involved. It is used to add security and blockchain features, such as immutability and asset management. This provides a mechanism of distributed database eventually adding the concepts of immutability, creation and movement of digital assets database. This is used in real time applications and it is perfect for the proposed work. Hence, a blend of the IoT and Blockchain is used to achieve a system for automated and secured storage and operation. The system is in turn enhanced with the user interface for customer operation. Ethereum and Node js are used for the front-end to develop a web app for access to documents and monitor wallet transactions. JavaScript and node JS are used for interactions and running the blockchain. Python is used to connect the operations of RFID Arduino with the server running. A smooth transaction between the hardware and software is achieved by using additional frameworks, serial packages and APIs.

5.1 Workflow of the Proposed System

To develop the proposed system, a number of frameworks are used to fit in the various processes and assemble them. The proposed system is portrayed below:

Step 1: A first time user registers their vehicle with a unique RFID by installing the RFID tag on the vehicle.

Step 2: The user creates an account on the web application with their credentials. The user adds wallet money. Minimum wallet money is needed for any parking area.

Step 3: When a vehicle arrives at a parking location, the RFID Reader at the location identifies the vehicle based on its unique ID and the time stamp is noted.

Step 4: When the vehicle leaves the location, the reader scans the ID and the ID is matched against the database and a time stamp is noted again.

Step 5: Based on the two time stamps, as per the policies of the parking location, the amount is deducted from the wallet, and money will be credited to the contract.

Step 6: This transfer is through the blockchain and is secured by recomputed hash.

Step 7: The user can use the application to upload their documents through the UI. The user can also modify their profile. These documents are saved in the ledger securely.

Step 8: The user retrieves specific or all the documents based on the requirement from the web application.

5.2 *Working of Smart Contracts*

Step 1: Creating a smart contract using Ethereum Solidity language.

Step 2: The Parking Area is the contract name and the parameters are Minimum Contribution, Add Users, Add Vehicle Number, Address, Vehicle details, Balance amount and Summary.

Step 3: Once all the details are stored then it will show the summary of a particular user.

Step 4: Once the vehicle leaves the parking area, and based on time stamp, money will be deducted from the wallet.

5.3 *Deploying Smart Contract into Blockchain:*

Step 1: Create an account in MetaMask. MetaMask is a distributed browser, where users can allow running EthereumdApp and browser. It is a blockchain transaction wallet that has a very secure identity vault.

Step 2: Once the MetaMask account is created, a secure key will be provided that should paste to Atom framework where smart contracts are running and the key should be pasted in the smart contract deploy file, which is in Json format.

Step 3: Create the Rinkeby Faucet network. This network is based on Ethererum for funding or accumulating enough ether. This network is used to prevent malicious actors or long running spam attacks. Three different types of ethers are provided:

3 ethers/8 h,
7.5 ethers/1 days,
18.75 ethers/3 days.

Step 4: After creating an account, the user will get an address from MetaMask. Copy MetaMask Address to any protected websites, such as, Google plus, Facebook, Twitter, etc., and get the web link of address.

Step 5: Now when requesting a fund through Google Plus then the MetaMask address should publish a new post with the Ethereum address embedded into the content and select the ether from the network.

Step 6: Create an account in Infura. Infura is also an Ethereum node cluster. It is hosted such that the user runs their own applications without being required to set up their own node or wallet.

Step 7: Once the infura account is created, then one has to create a new project. An infura link will be obtained, and it should be copied and pasted on the smart contract deploy file, which is in Json format.

Step 8: Once the Rinkeby transaction is over then MetaMask will get credited based on how much ether the user selected.

Step 9: We will have two Json files, one is compile.js and deploy.js. Once compile.js file is run, it will build a bytecode file, which will communicate with other networks. Once the compile.js file runs successfully, then the deploy.js file should be run, there it will show from address and attempting to some address.

Step 10: Once it runs successfully then it can check whether a smart contract is deployed or not. In MetaMask we can get an option like is seen in blockchain deployment using EthereumScan. Opening EthereumScan will redirect to the smart contract with from address and to address and time of deployment.

6 Result and Discussion

Each vehicle is installed with an RFID chip with a unique ID. Upon arrival of vehicle, the unique ID is detected and recorded. Arduino software is used for detection of vehicle. This will help to connect with python for storage of unique ID for future use. PLX-DAQ is used for live monitoring of data. The protocol used for communication is called Serial Peripheral Interface (SPI). SPI is a synchronous protocol for communication between devices connected with Arduino within a short distance. In this Work, AT Mega 2560 Arduino UNO is used for implementation of the IoT part. In order to record the data read from the RFID Chip on a Windows system, Arduino active port is connected via python code to write the contents into a file. Serial package in python is used to connect to the active port, COM4 with a baud rate of 9600. An output file is created and the unique IDs are appended.

6.1 Using Blockchain for Secured Transaction

Node.js is used for running the backend blockchain code. The RFID ID is detected and read from the file using file handling mechanisms of node.js. The unique IDs are in turn added as a link to the existing blockchain. This is used for secured transactions using hashing function. The first block is called the genesis block, which is the initial block. The second block is a unique vehicle with an ID, and the third block is of the parking area with its ID. A Time stamp is calculated to measure the amount of time the vehicle is parked in a particular area to facilitate fee collection. Hash value is calculated each time and transaction is changed.

6.2 File Storage

Public ledger is used for secured storage of documents, such as insurance papers, driver's license and emission test card. Two rows have been added, one each for a location and a vehicle. It can be viewed that the field added with the primary keys has the RFID numbers and a hash value is generated. Every user is provided a valid username and password as part of credentials. The login page has been designed to facilitate authorization of data. The username and password are matched against the saved database and only upon valid entry of username and password is a user allowed access. Session management has been taken care of to avoid multiple user access to the same account and time out. The aim of the work is to design a system to avoid traffic congestion and unnecessary waiting for parking. This also includes secured e-storage of important vehicle documents, such as driver's license, vehicle insurance and registration card on database. The concept of blockchain is used for achieving secured transactions at parking locations. The first module focuses on automatic vehicle identification. As an interesting concept of the IoT, RFID is used for the detection and identification of a vehicle based on its unique ID. Arduino AT Mega with packages MRFD522 and SPI are used for contactless communication between the vehicle and the location. The detected unique ID along with the time stamp is saved in a file. File Handling capabilities of Arduino are enhanced by the use of PLX-DAQ for live data monitoring into Excel. Python serial package is used for communication between the Arduino and the inbuilt system file manager on the serial communication port to save the file documents.

This feature is highly specific to Windows systems as on Linux systems the file can be saved without the intervention of Python. At the exit, the time stamp is collected again and based on the time interval a transaction of money is placed between the vehicles unique ID and the location unique ID. Node.js is used for running the blockchain for secured transactions. Unique IDs of the vehicles are used as keys of the blockchain. With every transaction, the values within the blocks are changed and this information is propagated to each block in the chain. New hash values are computed on all blocks to maintain integrity and consistency. The second module of blockchain is an integral part of secure money transfer and insurance monitoring. Based on the valid date, the documents are monitored for continuous expiry and intimated to the user beforehand and an order is placed with the vendor. The third module deals with the storage of important documents securely. Since the work is aimed at a decentralized mechanism and real time, a public ledger powered with a database is used for document storage. This provides the features of decentralization, scalability and security. The documents are hashed and stored in the database, which can be accessed as per need by the user using their profile credentials. To facilitate seamless interaction between the modules, serial port communication is achieved. For the benefit of the users who range from naive to expert, a user friendly and creative UI has been designed using the Ethereum platform. Thus, ultimately the aim of the work to integrate the IoT and Blockchain is achieved by incorporating decentralization, peer-to-peer communication, security, privacy, integrity and ease of use. The time of operation can be greatly reduced over the existing traditional parking systems.

Another research finding in the proposed system is transaction timing. Time is one of the major parts of human life. Everyone is busy and they do not want to waste a single minute. Here, a blockchain transaction is carried out very fast and automatically, whereas manually one would need to wait for payment.

7 Conclusion

The principal issues in most of the metro cities are due to the increasing number of vehicles and traffic congestion occurred in malls, public roads and parking lots. In this prevailing situation, a technological automation to monitor and maintain the parking and traffic is the need of the hour and can be achieved by integrating the features of blockchain and the IoT for secured transactions and cashless transfer without any delay or queuing. Such a solution would be a part of the Digital India Movement by reducing usage of physical currency notes and promoting the ecosystem by issuing e-receipts. With the ever increasing number of vehicles on the road today, a remarkable solution should be focused on helping to reduce traffic time and ensure easy parking. This work mainly focuses on helping to reduce waiting time due to traffic at various locations, including shopping malls, movie theatres, bus stands, railway stations, etc. Adding to these benefits is replacing the old hardcopy system with digital media and storage. Citizens need to possess their personal documents, including vehicle insurance, driver's license and other importance ID proofs. Saving a digital copy of the same with a secured authentication mechanism not only helps to reduce risks of losing the documents but also to produce it at any instance anywhere with just user identification. Every user must create their own wallet in the system and alway maintain a minimum amount in the wallet. A transaction is carried out automatically from the wallet once the user leaves the parking lot, and thus they do not have to wait for any queue and the system automatically generates e-receipts. Hence, the proposal emphasizes the integration of blockchain and the IoT for solving traffic issues and road transportation issues in the real world.

References

1. Ashton, K. (2009). That 'internet of things' thing. *RFID Journal, 22*(7), 97–114.
2. Nakamoto, S. (2008). *Bitcoin: A peer-to-peer electronic cash system.*
3. Davidsson, P., & Wiklund, J. (2007). *Levels of analysis in entrepreneurship research: Current research practice and suggestions for the future. Entrepreneurship* (pp. 245–265). Berlin: Springer.
4. Hossain, S. (2013). 5G wireless communication systems. *American Journal of Engineering Research (AJER), 2*(10), 344–353.
5. Shaikh, F. I., Jadhav, P. N., Bandarka, S. P., Kulkarni, O. P., & Shardoo, N. B. (2016). Smart parking system based on embedded system and sensor network. *International Journal of Computer Applications, 140*(12), 45–51.
6. Reddy, P. D., RajeshwarRao, A., & Ahmed, S. M. (2013). An Intelligent Parking Guidance and Information System by using image processing technique. *IJARCCE, 2*(10).
7. Patel, C., Swami, M., Saxena, P., & Shah, S. (2015). Rotary Automated Car Parking System. *International Journal of Engineering Science and Innovative Technology (IJESIT), 4*(2), 408–415.

8. Umar, W., Alam, F., & Usman Ali Shah, S. M. (2014) Optimized design of a Parking Management System using FPGA. In *2014 International Conference on Open Source Systems & Technologies*. IEEE
9. Gupta, A., Kulkarni, S., Jathar, V., Sharma, V., & Jain, N. (2017). Smart car parking management system using IoT. *American Journal of Science, Engineering and Technology, 2*, 112.
10. Gauravaram, P., Millan, W., Dawson, E., & Viswanathan, K. (2006). *Constructing secure hash functions by enhancing Merkle-Damgård construction. Australasian Conference on Information Security and Privacy*. Berlin: Springer.
11. Novo, O. (2018). Block chain meets IoT: Architecture for Scalable Access Management in IoT. *IEEE Internet of Things Journal, 5*, 1184.
12. Pala, Z., & Inanc, N. (2007). Smart parking applications using RFID technology. In *RFID Eurasia* (pp. 1–3).
13. McConaghy, T., Marques, R., Mller, A., De Jonghe, D., McConaghy, T., McMullen, G., et al. (2016). BigchainDB: A scalable blockchain database. White paper, BigChainDB.
14. Sharma, P. K., Moon, S. Y., & Park, J. H. (2017). Block-VN: A distributed block chain based vehicular network architecture in smart City. *Journal of Information Processing Systems, 13*(1), 84.
15. Rathore, M. M., Ahmad, A., Paul, A., & Rho, S. (2016). Urban planning and building smart cities based on the internet of things using big data analytics. *Computer Networks, 101*, 63–80.

Edge Computing-Based Intrusion Detection System for Smart Cities Development Using IoT in Urban Areas

S. Vimal, A. Suresh, P. Subbulakshmi, S. Pradeepa, and M. Kaliappan

1 Introduction

Internet of Things (IoT)-based industry has had enormous growth in urban planning with its foundations in the development of the industries for chips, microcontroller kits, electronic gadgets, and telecommunication systems, forming industrial segments. Attacks and vulnerability issues are the main concerns in the IoT, such as network breaches. Vulnerability can be suspected using an intrusion detection system, which can encounter anomalies with the reliable services of IoT applications. Various deep learning models have been proposed, with the modeling scheme of learning and data from sensors forming the structure of the setup. The

S. Vimal
Department of IT, National Engineering College, Kovilpatti, Tamilnadu, India
e-mail: svimalphd@gmail.com

A. Suresh (✉)
Department of Computer Science and Engineering, Nehru Institute of Engineering and Technology, Coimbatore, Tamilnadu, India
e-mail: prisu6esh@yahoo.com

P. Subbulakshmi
Department of CSE, School of Computing Science, Hindustan Institute of Technology and Science, Chennai, Tamilnadu, India
e-mail: subbu.psk@gmail.com

S. Pradeepa
School of Computing, Sastra University, Thanjavur, Tamilnadu, India
e-mail: pradeepa.pradee@it.sastra.edu

M. Kaliappan
Department of CSE, Ramco Institute of Technology, Rajapalayam, Tamilnadu, India
e-mail: kalsrajan@yahoo.co.in

G. R. Kanagachidambaresan et al. (eds.), *Internet of Things in Smart Technologies for Sustainable Urban Development*, EAI/Springer Innovations in Communication and Computing, https://doi.org/10.1007/978-3-030-34328-6_14

Internet for the sharing of global communication and content established connected objects and smart devices with the support of the International Data Corporation (IDC) [1]. The IoT paved the way for smart cities by supporting optimization and enhancement of public services in a smarter way, leading to smart transportation, smart parking, smart hospitalization, and urban development. The city offers a strategy of control over various domains established with the new service rule, with automatic classification of the services offered. Data from different sources have been collected, together with particular locations and times [2]. The expansion of cities is handled in a smarter way with smarter development in a cloud-based environment.

In an edge computing-based cloud setup, IoT data gathering includes all data to be gathered from the remote server. The remote data communication is established with the increased speed in the bandwidth. Fog computing and data processing have been done from remote servers, which increases the benefits of the data size and the low latency simulation in the filtered server data. Edge computing offers very large offloading and the storage accuracy has been enhanced with privacy preservation in a smart method of IoT prediction [3].

In the modern era, people are getting connected to the real world through the online sector, using an integrated circuit technology, and wireless communication has been established with signal acquisition and data preprocessing with wireless communication capabilities. The IoT has been divided into three layers: the perception layer, the network layer, and the application layer [4]. The perception layer deals with the end devices and tail nodes. The end devices monitor a progressive approach in elaborating the segment with the layer channel in the devices to support radiofrequency identification (RFID) tags, cameras, and GPS devices. The sensors are used by the parameters to form an environment in the sensing layer with the information from the physical world using analog-to-digital conversion [5]. The second phase of the IoT deals with the network layer, which connects the sensing layer with the application layer, including mobile communication and fixed communication using internet and private communication networks. The network layer includes various functionalities including optical fiber communication, security access, and satellite access [6]. The network layer deals with the applications in the application layer; human–computer interaction is one of the biggest applications to retrieve data from the valuable environment. End devices include IoT data retrieval from the environment within an information security and business domain. The security includes analysis of data, computers, storage, and various technologies. Data are transmitted between the end devices and the sensors [7].

The IoT brings a solution to a problem prevailing in the recent applications with problems in recent surveys of the existing RFID technology. In the next 5 years, the usage of the IoT is assumed to exceed 30 billion people [8]. RFID occupies a big place in the recent wireless communication paradigm. The security of the IoT includes authentication with a secure hash key for intrusion detection and external data behavior using an intrusion detection system (IDS). IoT security enables the system to enhance the hotspot of the system. Figure 1 shows the layered structure of the IoT in a Wi-Fi management system [9, 10].

The prospects of the IoT are enhanced in the development of IoT at all levels of industry. A survey indicates that China is developing an economy with the infrastructure to expand the country [11]. The rise of the IoT has huge involvement in smart homes, with the wireless local area networks (WLAN) dealing with the public market. Intruder detection involves faster intrusion with cost benefits. Various long-distance communications have been established with the various remote sensing assumptions in a sensing using the sensors in UART, Lora, Zigbee, and Bluetooth applications [12]. Ubiquitous Wi-Fi includes widespread application in IoT security-based technology and also the security limitations monitor the use of security with a hotspot in the current IoT [13]. The economic deployment has the option of considering data security, with deep learning algorithms achieving a remarkable role in IoT considerations. Deep learning involves data sampling that is satisfied in the real world. Deep learning is achieved in that the problem with the target data has been assumed in the good development of the deep neural networks [14]. The IoT in China has its own development with various research universities combining to be the university to communicate the performance in the classification of RFID research and development [15]. Image classification depends on the industry in the federation of China having a migration algorithm with the parameter to be analyzed. The layered architecture is represented in Fig. 1.

The security technology has authentication technologies to provide secure transmission with the passive defenses, with the state, behavior, and usage being provided with IoT security in devices [16]. The function of the IoT as a WLAN has a convenient environment that can automatically provide humidity and adjust the temperature, the IoT has the ability to monitor the temperature for Smart home management system [17]. Wireless technology has a big impact in sensing

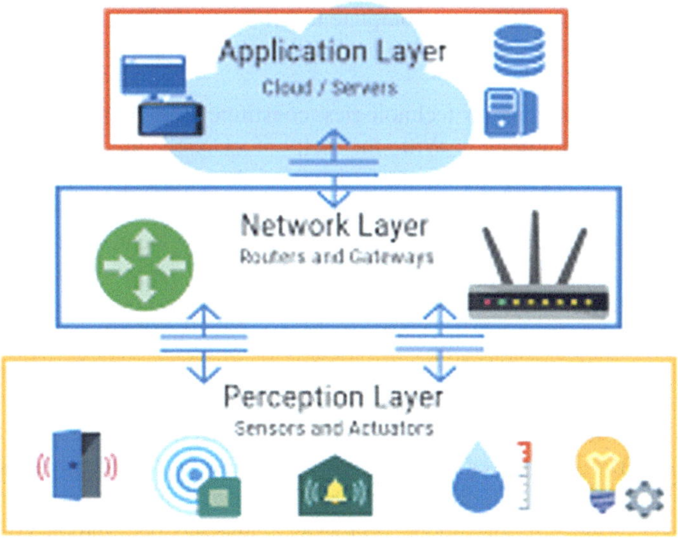

Fig. 1 Layered architecture supporting the Internet of Things

WiFi signals for IoT. The information sensed can be assumed to be additional communication that can be done in the establishment of Wi-Fi [18].

The migration of the learning categories to the essential methods of treating data with the inductive learning has the ability to analyse the abnormal data [19]. The dimensions of the training data and the test data have to be assumed to fit the marginal data. The inverse of the homogeneity problems, the regression of the data, can be directly assessed using the existing research. Direct push learning is one of the techniques in supervised learning and is effectively addressed in the tasks of unauthorized learning with clear characteristics. Intrusion is a process of unauthorised access and tampers the data with in the system [20]. The safety of resources and the sensors is in addition to the protection of the physical sensors. The first line of defense uses data encryption and sensor network security, key management, and the protocols have been identified by unauthorized parties [21].

The IoT sensor has attacker nodes and the weak points of the system include the fact that the IoT sensor has the nodes with a defensive nature [22]. The IoT has the best defined approaches for the defense system with the attacker being established in the automatic mechanism in the sensor network. The IDS has the best feature in the detection in the source/sink, with the nodes having a secure environment [23]. It uses an automated system for the attack behavior, which has low traffic in short wireless communication and an alarm signal within the node-based distributed architecture [24].

In the distributed environment of the IoT, the IDS collaborates with the IoT and has specific tasks in the specific attacks in the predetermined area of the IDS. The IoT helps to discover various attacks within the sensor nodes with the generation of agents in a communication activity in a specific location [25]. IDS agents are located within the neighboring node within the wireless communication channels and are profound in a behavioral norm in separate knowledge between areas in the foundation of better learning in a complete target field with the favor of the specific locations. The learning tasks are allocated with greater expression within the target field [26].

Mobile and smart phone technologies constitute one of the modern world communication technologies. With the help of sensors, the context has been extracted [27]. A recent technological survey on Android mobile phones needs a proximity context that senses the object in closer observation of the screen [28]. The android application along with the GPS predicts the location of the user. The recommender system is one of the process contexts in the information to predict the user prevalence in the relevant information. The information is assumed to be off in the tourist location prediction in the peer-to-peer information in the context extracted within the ratings of software and hardware features to a higher level with the values of converting GPS with the data information.

Research studies are concerned with context management in the predetermined applications. The low power processor is assumed to be to a continuous context aware in the proposed solutions in the accelerometer with the help of a mobile sensor. In everyday life the mobile user's behaviour has been predicted with the random number generator using the sensor-based mobile resources and the resource estimation is predicted easily [29]. The power booster system is assumed to be for the adaptation of mobile monitoring in the system in mobile applications.

2 Intrusion Detection System

Intrusion is an unauthorised activity by authenticated users and violation of the right to deliberately access by users to destroy and tamper with the data in the system. The IDS was initially brought into existence in 1980 by James P. Anderson. In an IDS, the key nodes are monitored in integrity with the nodes of the network and the collected information is transmitted to the monitorable key to improve confidentiality. The network is a collective resource as information gathering in a centralised place. The safety of the sensors is monitored in a consistent way to address the defense sensor network security in unauthorized parties with the support of the attacker. The IDS setup is used to detect the attacker and collectively achieve the intrusion to cooperate in the IoT sensor nodes. When the attacker is identified in the network, it automatically sends a signal to the transmitter to enable the assumption in the predefined network [30]. The attack behavior has low traffic and wireless communication has a code of vulnerable actions in the node corresponding in the IoT with the IDS. The IoT architecture has a client in the fashionable way, where each module develops an IDS system with most of the adoption in the required way of prediction in formation of the general cluster. The network forms a cluster and the IDS has the major application in occupying the cluster and monitoring it. Each cluster does the data routing and data management in the cluster to avoid the malicious packets in the wireless port [31].

The independent IDS has a sensing layer with the power supply based on need, and the installation in some other nodes outperforms the key nodes in the IoT using the perception layer. The nodes in the IoT work independently using data acquisition and achieves intrusion detection. It needs the information for the central node to acquire the detailed information in anomaly detection with the various modules in the system. The sensor nodes have limited computing and communication capabilities [32]. The nearby nodes in the unit allow the sensor to ensure that the intrusion occurred in the unit using the network capabilities. The intrusion behaviour is predicted and the sensing mechanism is very much supported with the mobile edge computing paradigm. Mutual cooperation may be established using the big data sensing platform [33].

3 IDS in RFID

An RFID system has electronic tags and readers to support the database server using frequency technology with the read and write electronic information in a tag control with the composition of the antenna and read and write module observation in the data processing module in a systematic way [34]. The reader sends the information to the data techniques and the mere processing of the applications can be achieved using the device controller. The exchange of information has been established with the transmission power and volume of the reader in an entire RFID system using electronic tags. The response system in the RFID is assumed to be of the electromagnetic wave signal using the system wave signal processing with the reader in the antenna system [35]. The reader system sends information in the electromagnetic wave using transmission with a sender report. The tag includes energy from the transmitted electromagnetic wave signal from the secure response and the mechanism energy is based on the tag with the reader information. The signal transmission and reception through the antenna are secure [36]. The reader fills the data reader and transmits the responder in the data segment in the area of response in the antenna.

The signal transmission has the reception with the encoded information, which has been given to the continuous carrier in the signal to tag by modulation in the radiofrequency (RF) signal. The tag modulates the RF signal with the support of the received information in the reflected coefficient antenna in the modulated schemes in support of the tag [37]. The probabilistic information has been attained in the concerned in the electromagnetic field (EMF) in the field region in the segmented approach. The reader has alternating control over the deposited information in the self-charge mode of capacity modulation [38]. The alternating EMF in free air has the space to inculcate the operating voltage within which the attached method is closely connected with the reader to produce inductance with recognition within a range of 10 cm [39].

The EMF has a better coupling method in an antenna with the reader, which sends a tag instruction to the load of the changing impedance in the tag, matching criteria for its own load appearance. The required information is matched by its own changing load impedance [40]. The tag includes required information in the data bits of the continuous stream. The EMF signal changes the state in a multi-aspect way of increasing the range from 15 to 20 m. The RFID system can scan multiple objects in the information and the tags are used to read the data in the format of the RFID tag to approach the reader/writer information in the secured manner of the environment. The RFID tag has been set with the EMF tag in the used environment with the desired location and sufficient needs.

Security is the biggest challenge in the detection of thefts and day by day it is emerging in urban areas owing to a lack of safety and security in commercial buildings and in factories. The traditional technologies are working on global systems for mobile communications (GSM) to increase home security to a certain level with the support of IoT devices using different sensor systems. The central controllers are being used to handle the different movements in the sensor system in the era of emerging technologies. The presence of humans in the IoT-based system uses a different system with the support of a microcontroller. This controller is used with different parameters in different devices. The home security system uses a very expensive and ineffective system with different situations and cannot fulfill the current prevailing security needs. The security system should be cost-effective with strong secured system that should fulfill the customer requirements. The changing needs to be addressed in the current scenario with the user are remotely accessed in the need to progress in home appliances for the development of wireless technology to the customer through the short messaging service (SMS).

4 Edge-Based IoT

The traditional IoT has the sensor data collection and information processing through the cloud server where similar scalability issues are monitored with the network, it has refined data in the scalability issues in the refined condition, with a separate module of the implementation with the draft of assuming the condition in the prescribed section and the technique to be assumed of the performance enhancement in the assumed transcription.

The technique of the IoT is to process the information with the offered needs in the environment and the multi-interface needs in the process of adding the wireless sensor networks (WSNs) to assimilate control in the cellular base station. The WSN has a needs-based approach in the edge server, having a cellular network with the control plane offering data-based control in accordance with the needs of the network. In the data sent to the network in the WSN the edge monitoring has to be accomplished with the cellular network coverage in the prescribed assumptions [41]. The edge server is one of the chosen cellular base stations in the network, which consists of the connections of the IoT device with data volatility.

Edge computing supports traditional cloud computing and fog computing with a desire to focus on recent articles that investigate various types of fog computing with different IoT applications. The IoT challenges are resolved using edge-connecting devices and hence the data segment may be achieved during the various IoT applications in future research directions of IoT metrics. More of a paradigm shift will be made in the analysis of the future of IoT, with various segments in the metrics for evaluating future research analysis in Fog based IoT.

The edge supports fog-based computing with the building blocks of the cloud set-up, as shown in Fig. 2. The fog can be associated with the edge in terms of characteristics:

- Location prediction and the decreased latency of the edge nodes can be made available in different locations and the fog is made available to the end devices. The latency may be automatically reduced in the fog-based resource allocation.
- The distribution is assumed to be off in geographical in nature and the centralized cloud is assumed to be deployed anywhere with fog services.
- Provides scalability and henceforth large number of sensors have been deployed and connected throughout the large-scale end devices.
- Mobility is done with the support of the frequently similar data processing and the location-based ID separation protocol is used for location-based services.

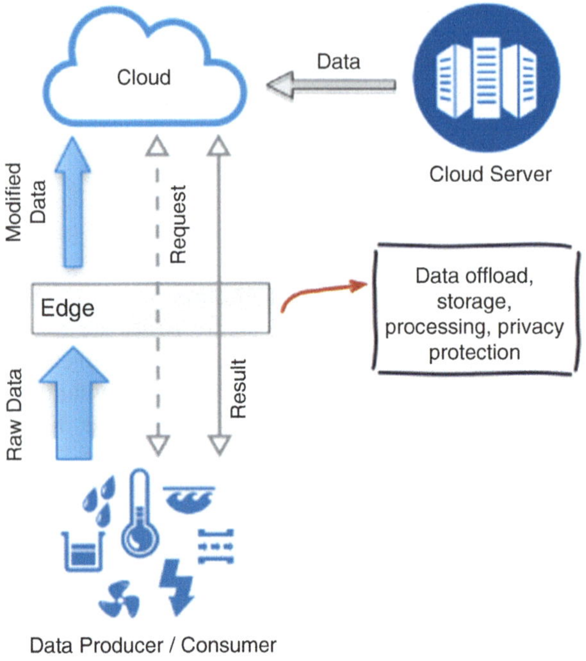

Fig. 2 Edge computing paradigm

- Close real-time interactions where fog computing provides interaction with other different fog applications and fog devices in cloud deployment in sensor-based edge computing.
- Heterogeneous in nature where the end nodes are from the different manufacturers in the cloud with susceptible deployment.
- The fog has the capability to provide the network in the nature of evolving the segmented network in the visible path of the different platforms.
- Interoperable in nature and it has fog based computing with the online support [48].
- Cloud analytics with edge connectivity has been real-time processing of data, which has been used with the end devices in the network. The different characteristics for selecting IoT architecture are discussed in Table 1.

5 IoT Edge Devices: Cloud Middleware

Internet of Things edge computing provides a detailed view of the cloud middleware in the processing of the IoT. IoT cloud middleware has various modules consisting of cloud controller, distributed message broker, and subscribers. The data management services are performed by the IoT cloud connectivity based on the surprising

Table 1 Characteristics for selecting IoT architecture

Table	Centralized architecture	Distributed architecture
Devices	Low number of devices can be connected	High number of devices can be connected
Frequency rate	Low frequency and the rate of sending the data is low	High frequency and rate of sending the data is high
Resource allocation	Higher	Economical
Rate of response	Lower	Higher

event of the raised module in the fixed data segmentation and it has been insisted upon to follow the distributed broker services with the message route between the subscribers and the IoT. The IoT device has been established with the devices to transform metadata into IoT data in the devices so that the filtered messages can be established with the NoSQL databases [42]. The controller sends and receives messages with the IoT data and Apache Cassandra has been used for storage management. The IoT data management is assumed to be in the NoSQL database and MongoDB has also been used with Hadoop Distributed Files System (HDFS) cloud storage. Dropbox is a form of storage freeware that has been used to fix multiple server usage in data streaming for processing the manufacturers to be in the architectural Simple Object Access Protocol (SOAP)-based interoperable scheme [47]. The devices that are assumed to be in the apparent processing in multiple servers have been in the final IoT with SOAP-based streaming in the interoperable IoT [43]. The message broker is allotted the range limit for the message received and transferred in the IoT section for the most asynchronous mode of the GET/POST method of defining the consistent view of predefined clients. This is a string model repository for the assumption of data on the subscribers in the method of pulling the alerted signals for custom service [51]. The IOT cloud controller is discussed in Fig. 3.

The custom service can be a user or a device that has a source to monitor the cloud controller in the noted single software program for the face detection mechanism with the sensor in the Face video interaction in the single mode of detection for the performance of the discover and control of a single software program having a sensor and it is also used to retrain the similar images. The protocol stack for forming the IoT is shown in Fig. 4.

Nowadays, the IoT has been emerged as the biggest contribution in the development of smart city. IoT-based smart cities have been implemented with the number of benefits and smart parking has been launched as a part of smart city development. The set of things has been monitored with the abstraction of a pattern in the observations in the sensor network. The approach has been considered in the expansion of cities, with many circumstances in the sensor network having multiple approaches to the decision-making principles in IoT coupled with the cloud infrastructures [44]. The timely decision-making of the projects has been done with the IoT within the cloud environment. Cloud processing has been done with the remote sensor data, which has more processing speed, and edge computing focuses on the use of the bandwidth and the processing speed. The storage has been sent

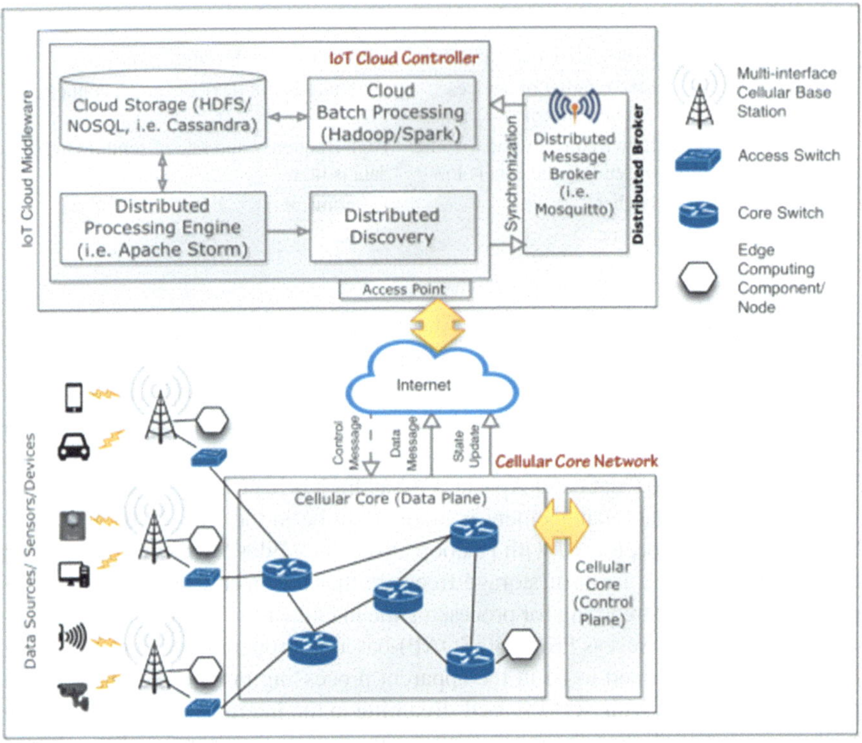

Fig. 3 Internet of Things-based cloud controller

Protocol Stack:	Application	Application	Application	Application
				CoAP
	Transport **TCP/UDP**	UDP / ICMP	TCP/UDP	UDP
	Internet **IP**	IPv6	**RPL**	IPv6/RPL
		6LoWPAN	6LoWPAN	6LoWPAN
	Network Access	IEEE 802.15.4	IEEE 802.15.4	IEEE 802.15.4
Typical scope:	WAN	PAN	PAN	PAN
Layer:	Internet/ Transport	Internet	Internet	Application
Name:	**TCP/IP**	**6LoWPAN**	**RPL**	**CoAP**

Fig. 4 Protocol stack for IoT architecture

to all the data management in the cloud with emerging data applications. The data is processed with the multiple speed and it may focus an edge computing-based paradigm to optimize the performance of the data processing in the network [45].

The data need to contact the remote server and hence the benefit of improving the large data sizes may occur and generates a low latency in the location awareness in the processing of data sources. Edge computing has a service in the remote server that significantly decreases in this method, with three benefits: large data sizes, it guarantees a low latency, and the location is based on the awareness of forming the location [49]. The edge is the closest processing of the data filtered in the cloud server with the option of finding data offloading, storage, processing, and privacy preservation in the network analytics. The smart IoT sensors has data that makes the decision-makers to be aware of the situation and it provides a reliable service to the smart urban development based on the edge mechanisms [50].

6 Distributed Processing

The portioning algorithm has been used in a distributed fashion for identifying the purpose of the mobile application in a partitioned way with the threading that migrates the use of the cloud, which executes the mobile integration part with the cloud application. The thread involves a new stack creation in the corresponding application using the invoking method in a new way of launching the threading condition in the Java virtual machine (JVM) [46]. Each method uses a module framework to store and retrieve the data in the corresponding manner of forming the condition in executable batchwise processing. A temporary clone has been created and may support the data for further processing to analyze using the IoT with edge computing to form the clustered data.

7 Cloudlet Computing in Edge-Based IoT

A datacenter ID formed to process a data structure with the cloudlet and the multicore computer are used with the cloudlet, which resembles a local store with a public or private area in VMware storage. Dropbox is open access with dynamic access in a synthesized and economical way and the VMs are employed in a separate multicore computer in a local store and dynamic VM synthesis is proposed in an efficient manner. The add-ons and the data in a Linux OS, which has an offloading option to excel in a similar aspect of enriching the content with the virtual box and the work related is assumed to be in the big data computing for smart cities development [52].

7.1 IDS in Urban Areas Using the Cloud-Based IoT

- In a normal system a loud blast sound is created, which is noticeable to the intruder(s) or creates a disturbance in the neighborhood, and this creates a panic situation.

- To avoid this, our system does not create any sounds and the videos are stored first and then retrieved later by the user; thus, it causes a delay, so that the user cannot be alerted suddenly.
- But in the proposed system, the video is sent to the cloud and can be retrieved at the same time by streaming; a text message is also sent to the user, so that the user can alert those nearby to stop the intruder.

7.2 System Architecture

This is aimed at presenting a home security system using Raspberry Pi to alert the owner by GSM message and a video clip to the owner's webpage. The proposed system attempts to see the video clip, featuring the intruder [13]. This system streams a video while any intrusion is happening, which is sent to the cloud and is viewed at the owner's webpage, and they also receive a GSM alert message. In this work, the system gives an alert regarding any intrusion that has occurred by means of GSM message and a video of the intrusion. This software model can be used if any visitors arrive; the Raspberry Pi continuously streams the video and sends it to the owner's web page through the cloud. This home security system could be good for use in many places such as banks, hospitals, laboratories, etc., and dramatically reduces the risk of unauthorized entry in both rural and urban areas [14].

7.3 IDS Using Edge Computing in Raspberry Pi

- Initialize the whole library to run the code in Raspberry Pi.
- Enable the Pi camera in the Preferences.
- Import the camera into the code.
- Define output pins (general-purpose input/output [GPIO] 18) for a passive infrared (PIR) sensor in Raspberry Pi.
- The setup() function is used to set the PIR sensor to sense the motion.
- If a motion or human heat is detected, the sensor triggers a signal to the camera.
- The camera.capture function is used to capture the video.
- The camera's start() function starts to record a video.
- The camera's stop() function stops the recording.
- The sleep() function gives a sleep time for the camera.
- The def SendMail() function authenticates and sends mail.
- The def mblesms() function triggers the Raspberry Pi to send an alert message to the user.
- The SendMail() function has shell commands that upload the video to the cloud.

7.4 *Algorithm Description*

The Raspberry Pi 3 microcontroller is first initialized using the Raspbian operating system. The Raspbian operating system works on Python programming. The Pi camera is initialized by enabling the camera in the preferences, and then installing and updating the required library files. Then, the required files are imported and the camera and PIR sensor are defined by declaring the output GPIO pin of the sensor [15]. After that, the function to detect the output of the sensor is defined: if the sensor gives a high output, then the camera function is triggered and it starts to capture with the start capture() function and captures until the time to sleep. Then an alert mail is sent to the user using the sendmail() function, and the sendmessage() function is also triggered and an alert message is sent. The sendmail() function has a shell that uploads the video to the cloud and can be retrieved using an Android application.

8 System Model

The proposed model consists of three modules:

- Microcontroller module
- Mobile application module
- Database module

8.1 *Microcontroller Module*

The microcontroller module has three parts:

- Raspberry Pi 3
- PIR sensor
- Pi camera

The quad-core Raspberry Pi 3 is both faster and more capable than its predecessor, the Raspberry Pi 2. For those interested in benchmarks, the Pi 3's CPU—the board's main processor—has roughly 50–60% better performance in 32-bit mode than that of the Pi 2, and is 10× faster than the original single-core Raspberry Pi (based on a multi-threaded CPU benchmark in SysBench). Compared with the original Pi, real-world applications will see a performance increase of between 2.5×—for single-threaded applications—and more than 20×— when video playback is accelerated by the chip's NEON engine [16]. Although the Pi can run many different operating systems, if stability and performance are required, then the official Raspbian operating system is a good choice, having been tuned to get the most from the Pi, and bundling a fast web browser and a good selection of office and programming software.

The passive infrared sensor determines the heat and motion of the human and sends a signal to the Pi camera. PIRs are basically made of a pyroelectric sensor (which can be Fig. 5 seen below as a round metal can with a rectangular crystal in the center), which can detect levels of infrared radiation. Everything emits a low level of radiation, and the hotter something is, the more radiation is emitted. The sensor in a motion detector is actually split into two halves. The reason for this is that we are looking to detect motion (change), not average IR levels. The two halves are wired up so that they cancel each other out. If one half sees more or less IR radiation than the other, the output will swing high or low. The PIR sensor itself has two slots in it; each slot is made of a special material that is sensitive to IR. The lens used here is not really doing much and so we see that the two slots can "see" out past some distance (basically the sensitivity of the sensor). When the sensor is idle, both slots detect the same amount of IR, the ambient amount radiated from the room or walls or outdoors. When a warm body such as a human or animal passes by, it first intercepts one half of the PIR sensor, which causes a positive differential change between the two halves. When the warm body leaves the sensing area, the reverse happens, whereby the sensor generates a negative differential change. These change pulses are what are detected. The Adafruit PIR has a trimpot on the back for adjusting sensitivity. This can be adjusted if the PIR is too sensitive or not sensitive enough—clockwise makes it more sensitive [53].

The Pi camera then captures photos and records video and sends it to the cloud. The camera module is a great accessory for the Raspberry Pi, allowing users to take still pictures and record video in full HD. It can capture images in low light. The Raspberry Pi Camera Board v2 is a high quality 8-megapixel Sony IMX219 image sensor custom-designed add-on board for Raspberry Pi, featuring a fixed focus lens. It is capable of 3,280 × 2,464-pixel static images, and also supports 1080p30, 720p60, and 640 × 480p90 video. It attaches to the Pi by way of one of the small sockets on the board's upper surface and uses the dedicated camera serial interface (CSi), designed especially for interfacing with cameras.

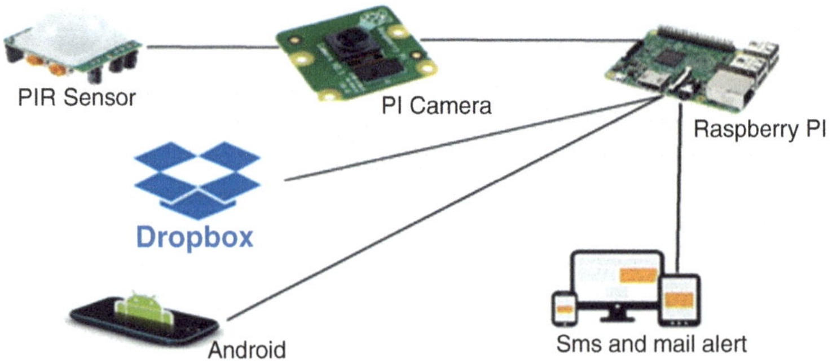

Fig. 5 Block diagram of the proposed system

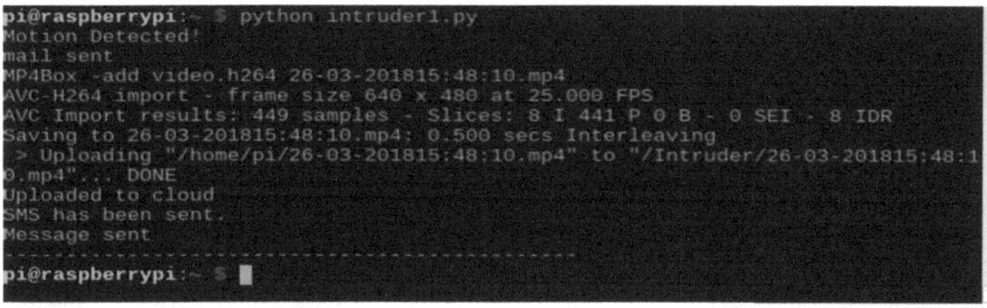

Fig. 6 Displays the execution of the program and uploaded video

8.2 *Mobile Application Module*

The captured video is stored in the cloud to retrieve the information. This Android application is installed on the user's mobile phone to retrieve the video. The application is designed in the Android studio, and uses an Android program to develop the main activity. The page has a list of the videos in the cloud. The main activity coding consists of the page design and the media player that is used to play the video.

8.3 *Database Module*

- A Dropbox database is used.
- It stores the data sensed by the Raspberry Pi.
- Once a file is added to Dropbox, it is synced to secure online servers. All files stored online by Dropbox are encrypted and kept in secure storage servers.
- The videos are stored with random names.
- The video can be on any format.
- The video can be of any size.
- The working model of the proposed system is shown in Fig. 5. There is a PIR sensor to detect human presence. The Pi camera is used to record video or capture images. The Raspberry Pi 3 kit is used as a microcontroller, and it runs in the Raspbian operating system. The intruder video is recorded and uploaded to the database.

The block diagram of our proposed system and the PIR sensor detects the presence of humans by heat and motion. Once the microcontroller identifies the intruder as shown in Fig. 6, the Pi camera captures the intruder and records a video, which it then stores in a database, as shown in Fig. 7. The Raspberry controller alerts the users by email and SMS and also sends the video to the Android application.

Fig. 7 Captured video using Android

9 Conclusion and Future Work

In this proposal, different technology from that that made urban areas prevail in the IoT-enabled edge computing and IDSs are discussed. An application has been described showing that in the event of any visitors arriving, the Raspberry Pi continuously streams the video, stores it in the cloud, and sends it to the owner's device. It can already be programmed and controlled remotely through a smartphone to do so much. Many cameras can be programmed to automatically record or take photos when motion is detected, and immediately notifies the user via text message or email. Through a smartphone, the user can communicate with an intruder through an intercom, turn the lights on or off, and even view a live camera feed. It could be for a good application for many places, such as banks, hospitals, laboratories, etc., and dramatically reduces the risk of unauthorized entry in both rural and urban areas.

References

1. Arab News. (2016). *30% increase in Umrah Pilgrims Seen by 2020*. Retrieved March 26, 2018, from http://www.arabnews.com/node/936801/saudi-arabia
2. Aazam, M., & Huh, E. (2015). Fog computing micro datacenter based dynamic resource estimation and pricing model for IoT. In *Proceedings of the IEEE 29th International Conference on Advanced Information Networking and Applications*, Guwangiu (pp. 687–694).
3. Aazam, M., Khan, I., Alsaffar, A., & Huh, E. (2014). Cloud of things: Integrating internet of things and cloud computing and the issues involved. In *Proceedings of the 11th International Bhurban Conference on Applied Sciences & Technology*, Islamabad (pp. 414–419).

4. Abdelwahab, S., Hamdaoui, B., Guizani, M., & Znati, T. (2016). Cloud of things for sensing-as-a-service: Architecture, algorithms, and use case. *IEEE Internet of Things Journal, 3*(6), 1099–1112.
5. Vimal, S., Kalaivani, L., Kaliappan, M., Suresh, A., Gao, X.-Z., & Varatharajan, R. (2018). Development of secured data transmission using machine learning based discrete time partial observed Markov model and energy optimization in Cognitive radio networks. *Neural Computing and Applications*. https://doi.org/10.1007/s00521-018-3788-3.
6. Vimal, S., Kalaivani, L., & Kaliappan, M. (2017). Collaborative approach on mitigating spectrum sensing data hijack attack and dynamic spectrum allocation based on CASG modeling in wireless cognitive radio networks. *Cluster Computing*.https://doi.org/10.1007/s10586-017-1092-0.
7. Abu Alsheikh, M., Jiao, Y., Niyato, D., Wang, P., Leong, D., & Han, Z. (2017). The accuracy-privacy tradeoff of mobile crowdsourcing. *IEEE Communications Magazine, 55*(6), 132–139.
8. Agca, R., & Özdemir, S. (2013). Foreign language vocabulary learning with mobile technologies. *Procedia—Social and Behavioral Sciences, 83*, 781–785.
9. Ilango, S. S., Vimal, S., Kaliappan, M., et al. (2018). Optimization using artificial bee Colony based clustering approach for big data. *Cluster Computing*. https://doi.org/10.1007/s10586-017-1571-3.
10. Ahmed, F. (2015). Comparative analysis for cloud based e-learning. *Procedia Computer Science, 65*, 368–376.
11. Ahmed, E., Yaqoob, I., Hashem, I., Khan, I., Ahmed, A., Imran, M., & Vasilakos, A. (2017). The role of big data analytics in internet of things. *Computer Networks, 129*, 459–471.
12. Akkaya, D. (2013). *Wireless inspirational bits for facilitating early design, degree project in communication systems, information and communication technology*. Stockholm: KTH Royal Institute of Technology.
13. Akyildiz, I., & Jornet, J. (2010). The internet of nano-things. *IEEE Wireless Communications, 17*(6), 58–63.
14. Akyildiz, I., Nie, S., Lin, S., & Chandrasekaran, M. (2016). 5G roadmap: 10 key enabling technologies. *Computer Networks, 106*, 17–48.
15. Alam, S., Chowdhury, M., Noll, J., & Senaa S. (2010). An event-driven sensor virtualization approach for internet of things cloud. In *Proceedings of the IEEE International Conference on Networked Embedded Systems for Enterprise Applications*, Suzhou, China.
16. Almarabeh, T., Majdalawi, Y., & Mohammad, H. (2016). Cloud computing of e-government. *Communications and Network, 8*, 1–8.
17. Alnanih, R., Ormandjieva, O., & Radhakrishnan, T. (2013). Context-based and rule-based adaptation of mobile user interfaces in mhealth. *Procedia Computer Science, 21*, 390–397.
18. Aloi, G., Caliciuri, G., Fortino, G., Gravina, R., Pace, P., Russo, W., & Savaglio, C. (2017). Enabling IoT interoperability through opportunistic smartphone-based mobile gateways. *Journal of Network and Computer Applications, 81*, 74–84.
19. Alvi, S., Shah, G., & Mahmood, W. (2015). Energy efficient green routing protocol for internet of multimedia things. In *Proceedings of the IEEE 10th international conference on intelligent sensors, Sensor Networks and Information Processing*, Singapore.
20. Amin, R., Kumar, N., Biswas, G., Iqbal, R., & Chang, V. (2018). A light weight authentication protocol for IoT-enabled devices in distributed cloud computing environment. *Future Generation Computer Systems, 78*, 1005–1019.
21. An, J., Gui, X., Yang, J., Sun, Y., & He, X. (2015). Mobile crowd sensing for internet of things: A credible crowdsourcing model in mobile-sense service. In *Proceedings of the IEEE international conference on multimedia big data*, Beijing, China (pp. 92–99).
22. Andersson, M. (2015). *Short range low power wireless devices and internet of things (IoT)*, White Paper UBX-14054570. U-blox.
23. Khan, Z., Anjum, A., & Kiani, S. L. (2013). Cloud based big data analytics for smart future cities. In *Proceedings of the IEEE/ACM 6th international conference on utility and cloud computing*, IEEE Computer Society (pp. 381–386).

24. Fan, W., & Bifet, A. (2013). Mining big data: Current status, and forecast to the future. *ACM SIGKDD Explore Newsletter, 14*(2), 1–5.
25. Bertot, J. C., & Choi, H. (2013). Big data and e-government: Issues, policies, and recommendations. In *Proceedings of the 14th annual international conference on digital government research*, ACM (pp. 1–10).
26. Kramers, A., Hojer, M., Lovehagen, N., & Wangel, J. (2014). Smart sustainable cities–exploring ICT solutions for reduced energy use in cities. *Environmental Modelling & Software, 56*, 52–62.
27. Kim, G. H., Trimi, S., & Chung, J. H. (2014). Big-data applications in the government sector. *Communications of the ACM, 57*(3), 78–85.
28. Tantatsanawong, P., Kawtrakul, A., & Lertwipatrakul, W. (2011). Enabling future education with smart services. In *SRII Global Conference (SRII), 2011 Annual IEEE* (pp. 550–556).
29. Neirotti, P., De Marco, A., Cagliano, A. C., Mangano, G., & Scorrano, F. (2014). Current trends in smart city initiatives: Some stylised facts. *Cities, 38*, 25–36.
30. Aguilera, G., Galan, J. L., Campos, J. C., & Rodriguez, P. (2013). An accelerated-time simulation for traffic flow in a smart city. *FEMTEC, 2013*, 26.
31. U.S. Department of Energy. *Smart grid/Department of Energy*. Retrieved September 23, 2015, from http://energy.gov/oe/technology-development/smart-grid
32. Yin, J., Sharma, P., Gorton, I., & Akyoli, B. (2013). Large-scale data challenges in future power grids. In *Service oriented system engineering (SOSE), 2013 IEEE 7th international symposium on IEEE* (pp. 324–328).
33. Mohamed, N., & Al-Jaroodi, J. (2014). Real-time big data analytics: Applications and challenges. In *2014 International Conference on High Performance Computing & Simulation (HPCS)* (pp. 305, 310).
34. Chen, T., et al. (2014). A sentence vector based over-sampling method for imbalanced emotion classification. In A. Gelbukh (Ed.), *Computational linguistics and intelligent text processing* (pp. 62–72). Berlin: Springer.
35. Labrinidis, A., & Jagadish, H. (2012). Challenges and opportunities with big data. *Proceedings of the VLDB Endowment, 5*(12), 2032–2033.
36. Bakshi, K. (2012). Considerations for big data: Architecture and approach. In *Proceedings of IEEE Aerospace Conference.*
37. Giese, M., et al. (2013). Scalable end-user access to big data. In *Big data computing* (pp. 205–245). Boca Raton: CRC Press.
38. Hashem, I. A. T., et al. (2015). The rise of "big data" on cloud computing: Review and open research issues. *Information Systems, 47*, 98–115.
39. Kannan, N., Sivasubramanian, S., Kaliappan, M., Vimal, S., & Suresh, A. (2018). Predictive big data analytic on demonetization data using support vector machine. *Cluster Computing.* https://doi.org/10.1007/s10586-018-2384-8.
40. Han, H., et al. (2014). Toward scalable systems for big data analytics: A technology tutorial. *IEEE Access, 2*, 652–687.
41. Buza, K., Nagy, G. I., & Nanopoulos, A. (2014). Storage-optimizing clustering algorithms for high-dimensional tick data. *Expert Systems with Applications, 41*(9), 4148–4157.
42. Geetha, R., Sivasubramanian, S., Kaliappan, M., et al. (2019). Cervical Cancer identification with synthetic minority oversampling technique and PCA analysis using random Forest classifier. *Journal of Medical System, 43*, 286. https://doi.org/10.1007/s10916-019-1402-6.
43. Dahan, H., Cohen, S., Rokach, L., & Maimon, O. (2014). *Proactive data mining using decision trees proactive data mining with decision trees* (pp. 21–33). New York, NY: Springer.
44. Suresh, A., Udendhran, R., & Vimal, S. (2019). An intelligent grid network based on cloud computing infrastructures, novel practices and trends in grid and cloud computing. *Novel Practices and Trends in Grid and Cloud Computing*, pp. 59–73. https://doi.org/10.4018/978-1-5225-9023-1.ch005.
45. Zhao, W., Ma, H., & He, Q. (2009). Parallel k-means clustering based on mapreduce. In *Cloud computing* (pp. 674–679). Berlin: Springer.

46. Suresh, A., Udendhran, R., & Vimal, S. (2019). Cloud-based predictive maintenance and machine monitoring for intelligent manufacturing for automobile industry. *Novel Practices and Trends in Grid and Cloud Computing*, pp. 74–81. https://doi.org/10.4018/978-1-5225-9023-1.ch006.
47. Taheri, J., et al. (2013). A bee colony based optimization approach for simultaneous job scheduling and data replication in grid environments. *Computers and Operations Research, 40*(6), 1564–1578.
48. Paul, A., et al. (2013). Video search and indexing with reinforcement agent for interactive multimedia services. *ACM Transactions on Embedded Computing Systems, 12*(2), 1–16.
49. Vimal, S., Khari, M., Dey, N., Crespo, R. G., & Robinson, Y. H. (2020). Enhanced resource allocation in mobile edge computing using reinforcement learning based MOACO algorithm for IIOT. *Computer Communications, 151*, 355–364.
50. Mariappan, D. E., Kaliappan, M., & Vimal, S. (2016). Energy efficient routing protocol using Grover's searching algorithm using MANET. *Asian Journal of Information Technology, 15*(24).
51. Dittrich, J., Blunschi, L., & Vaz Salles, M. (2011). Movies: Indexing moving objects by shooting index images. *GeoInformatica, 15*(4), 727–767.
52. Sun, D.-W., et al. (2012). Modeling a dynamic data replication strategy to increase system availability in cloud computing environments. *Journal of Computer Science and Technology, 27*(2), 256–272.
53. Vimal, S., et al. (2016). Secure data packet transmission in MANET using enhanced identity based cryptography. *International Journal of New Technologies in Science and Engineering, 3*(12), 35–42.

Index

G. R. Kanagachidambaresan et al. (eds.), *Internet of Things in Smart Technologies for Sustainable Urban Development*, EAI/Springer Innovations in Communication and Computing, https://doi.org/10.1007/978-3-030-34328-6

I

The manufacturer's authorised representative in the EU is Springer Nature Customer Service Centre GmbH, Europaplatz 3, 69115 Heidelberg, Germany. If you have any concerns regarding our products, please contact ProductSafety@springernature.com

Printed and bound by CPI Group (UK) Ltd, Croydon, CR0 4YY
05/10/2026
02228120-0003